Contract Documentation for Contractors

THIRD EDITION

VINCENT POWELL-SMITH
LLM, DLitt, FCIArb

JOHN SIMS
FRICS, FCIArb

and

CHRISTOPHER DANCASTER
DipICArb, FRICS, FCIArb

b

**Blackwell
Science**

Copyright © The Estate of Vincent Powell-Smith
and John Sims, 1985, 1990; new material ©
Christopher Dancaster 2000
Blackwell Science Ltd
Editorial Offices:
Osney Mead, Oxford OX2 0EL
25 John Street, London WC1N 2BL
23 Ainslie Place, Edinburgh EH3 6AJ
350 Main Street, Malden
  MA 02148 5018, USA
54 University Street, Carlton
  Victoria 3053, Australia
10, rue Casimir Delavigne
  75006 Paris, France

Other Editorial Offices:

Blackwell Wissenschafts-Verlag GmbH
Kurfürstendamm 57
10707 Berlin, Germany

Blackwell Science KK
MG Kodenmacho Building
7–10 Kodenmacho Nihombashi
Chuo-ku, Tokyo 104, Japan

First edition published by Collins Professional and
Technical Books 1985
Second edition published by BSP
Professional Books 1990
Third Edition published by
Blackwell Science 2000

Set in 10/13pt Times
by DP Photosetting, Aylesbury, Bucks
Printed and bound in Great Britain by
MPG Books Ltd, Bodmin, Cornwall

The Blackwell Science logo is a trade mark of
Blackwell Science Ltd, registered at the United
Kingdom Trade Marks Registry

DISTRIBUTORS

Marston Book Services Ltd
PO Box 269
Abingdon
Oxon OX14 4YN
(*Orders:* Tel: 01235 465500
         Fax: 01235 465555)

USA
Blackwell Science, Inc.
Commerce Place
350 Main Street
Malden, MA 02148 5018
(*Orders:* Tel: 800 759 6102
             781 388 8250
         Fax: 781 388 8255)

Canada
Login Brothers Book Company
324 Saulteaux Crescent
Winnipeg, Manitoba R3J 3T2
(*Orders:* Tel: 204 837-2987
         Fax: 204 837-3116)

Australia
Blackwell Science Pty Ltd
54 University Street
Carlton, Victoria 3053
(*Orders:* Tel: 03 9347 0300
         Fax: 03 9347 5001)

A catalogue record for this title is available from
the British Library

ISBN 0-632-05202-3

Library of Congress
Cataloging-in-Publication Data
Powell-Smith, Vincent.
    Contract documentation for contractors/
Vincent Powell-Smith, John Sims, and
Christopher Dancaster.—3rd ed.
      p. cm.
    Includes index.
    ISBN 0-632-05202-3 (hb)
    1. Contracts for work and labor—Great
Britain. 2. Contractors—Legal status, laws,
etc.—Great Britain. I. Sims, John, 1929—
II. Dancaster, Christopher. III. Title.

KD1638.P69 2000
346.4102′2—dc21
                                99-087085

For further information on
Blackwell Science, visit our website:
www.blackwell-science.com

Contents

Acknowledgements

Extracts from the NJCC Code of Procedure for Selective Tendering and the CIB Code of Practice are reproduced by kind permission of the copyright holders, the National Joint Consultative Committee, and its successor the Construction Industry Board.

Extracts from the JCT documentation are reproduced with the kind permission of the copyright owners, the Joint Contracts Tribunal.

The forms of application for the appointment of an arbitrator and the nomination of an adjudicator are reproduced by kind permission of the Royal Institution of Chartered Surveyors, the copyright holder.

The specimen agenda for a site meeting and the record of a site meeting are reproduced with amendments from *Contract Administration* (eighth edition, 1996), The Aqua Group, Blackwell Science.

From the Preface to the First Edition

Our object in writing this book is to provide contractors and their staff with *examples* of the sort of correspondence which they may have to write during the running of a contract under JCT 80 conditions. We have also given specimens of other contractual documentation. We would wish to emphasise that the examples are *guides* as to what should be written; it is impossible to standardise contractual correspondence and it would be dangerous to attempt to do so. The specimen letters are a guide and nothing more. They must not be followed slavishly. At various points contracts require that the contractor should send this notice or that, or else envisage that the contractor will write to the architect. In our experience, a great many contractors fall down on this vitally important aspect of contract administration, often to their own detriment.

We have not attempted to cover every situation which can arise in practice. Instead, we have concentrated on some of the more important aspects of contract documentation with which the contractor is concerned. In general, it is our view that, in letters and notices sent under the various contract terms, the actual wording of the contract should be followed so far as possible even if this leads to a certain verbosity and to pompous language. It may well be that, at times, tempers become frayed, but all extreme language and wild generalisations should be avoided in correspondence. It is better to stick to the facts, and to bear in mind that all contractual correspondence and documentation may one day be subject to scrutiny in litigation or arbitration.

Vincent Powell-Smith
John Sims
1985

Preface to the Third Edition

I jumped at the chance when offered the opportunity of preparing a third edition of this book.

To follow in the footsteps of such luminaries as Vincent Powell-Smith and John Sims is an honour indeed. I only hope that I can live up to their exacting standards. Vincent Powell-Smith, who wrote so many books, sadly passed away a couple of years ago. John Sims feels that his days as the author of books are over which leaves me in the hot seat. I shall always be grateful to John for affording me the opportunity of collaborating with him in the production of the second edition of *Construction Arbitration* which was my first faltering step on the rocky road of authorship. I must have done something right as Blackwells looked to me when John declined to do this third edition.

In my day to day practice as expert witness, adjudicator and arbitrator I spend a good deal of my time picking up the pieces of other people's problems. Before I got involved in this field I saw the possibility of having to get involved in a dispute which had to be resolved by an outside third party as a nightmare. Fortunately, in 30 years as a full time quantity surveyor acting for clients on both the employer and contractor sides of the industry I was able to avoid getting anywhere near arbitration or any other form of dispute resolution. Some may say that this was sheer luck. I prefer to think that it was the application of professionalism (with a small p) and being aware of the rules of the game that I was involved in.

My current involvement in dispute resolution allows me the privilege of seeing at first hand the mess that some people get themselves into. On many occasions it becomes quite evident that there has been absolutely no attempt to comply with the terms of the contract, that is, the rules of the game, and things have just gone from bad to worse. It often occurs to me that, with a little guidance, the participants in these contracts could so easily have avoided their problem. I hope that this book will continue to provide useful guidance of this nature.

This book is, as was said in the preface to both earlier editions, intended to provide contractors and their staff with *examples* of the sort of correspondence which they may have to write when dealing with the JCT forms of contract. The first edition of this book concentrated on the JCT 80 contract. The second edition widened this to cover IFC 84 and the JCT Minor Works contract. These earlier editions did not consider the other important JCT contract which has come into prominence, that is JCT With Contractor's Design, and for the first time, the book now takes account of this form. The JCT also produce a Prime Cost Contract, a Measured Term Contract and a Management Contract, and there is a multiplicity of other standard contracts in use. However it was decided that the book would become too complicated if it was widened beyond the four JCT forms that are principally used in the industry.

Since this new edition was planned new editions of all the JCT contracts have been published. We now have JCT 98 and all the other JCT forms have been similarly updated. In addition, following the Housing Grants, Construction and Regeneration Act 1996, we now have adjudication running up fast on the rails, and the book has been revised to take account of this and the new payment provisions introduced by the Act.

The wise words of the authors of the earlier editions are as valid now as when they were written. The Preface to the first edition should therefore be read as it sets out the *raison d'etre* of this book.

My thanks to Julia Burden and the others at Blackwells who actually make the final stages of the production of a book such as this a pleasure!

Christopher Dancaster
Milton Keynes
January 2000

Chapter 1
Introduction

1.01   General

A glance at any book dealing with construction contracts will identify the vast range of contracts that apply to construction works. There are contracts produced by the Joint Contracts Tribunal (JCT), the Government Contracts, contracts published by the Construction Confederation, the Institution of Civil Engineers, the Fédération Internationale des Ingénieurs Conseils (FIDIC) and the Civil Engineering Contractors' Association to name but a few. There are a number of other contracts that have not achieved a wide acceptance and also a propensity by certain employers and their legal advisers to produce one-off contracts for specific building projects.

Each of these contracts has its adherents and many of them are aimed at specialist applications but it cannot really be of assistance to the construction industry to have to deal with such a wide range of contracts. But that is the status quo, the proponents of each contract are not going to change easily and it is the contracting side of the business that has to adjust to fit as it tenders for and builds each project.

That said, the Standard Form of Building Contract (JCT 98, as it now is) is probably still the set of conditions most likely to be used for major building works where the design is carried out by a professional team retained throughout the project by the employer.

The other JCT contract that has come into far wider use in recent years is the With Contractor's Design contract (WCD 98). This is now, as with all the other JCT forms, in its 1998 Edition. This contract provides the employer with one point of responsibility. The end result is likely in that instance to be more cost than design led, and it is a matter of opinion whether that is a good or bad thing.

As always, the Minor Works contract (MW 98) remains the best seller of all the JCT contracts and the JCT Intermediate Form has its adherents. Both these contracts are in new editions being now known as MW 98 and IFC 98 respectively.

I have followed the precedent of the earlier editions of this book by writing it around the clauses of JCT 98 and indicating within each document the clause numbers from the JCT Intermediate Form (IFC 98), MW 98 and WCD 98 to which it will apply.

JCT 98 operates on the premise that there is an architect who designs and administers the project through to completion. IFC 98 and MW 98 use the term 'Architect/Contract Administrator' for the person fulfilling the same position. WCD 98 is written to the effect that the employer carries out the necessary administrative functions albeit he will probably engage the services of an employer's agent to act in his stead. In WCD 98 the design responsibility is, of course, with the contractor.

For the sake of simplicity I continue the practice of my predecessor authors and concentrate on the Standard Form (JCT 98) in the text. I generally use the word 'architect' throughout in order to avoid a tremendous amount of needless repetition.

One of the major innovations leading to the wholesale production of new editions is the amendment of dispute resolution procedures. One is due to the enactment of the Housing Grants, Construction and Regeneration Act 1996, or 'Construction Act' as it is often referred to in the building industry. This Act introduced the concept of a statutory right to adjudication to all construction contracts. The other is the decision of the JCT to provide the option of litigation as well as arbitration, as the contractual route for the resolution of disputes.

This last decision could well be a retrograde step. We have adjudication as the fire fighting process, the 'quick and dirty fix' as Lord Ackner described it in the House of Lords. This is, at the time of writing, beginning to be effective in freeing up cash flow in the industry and is taking the place of the notice of arbitration as the step taken to signify that the applicant is not prepared to be messed about any longer. The adjudication process will be in the hands of a technical specialist. Where better for the majority of disputes to end up but in the same hands? The adjudicator may well not have the time to get the answer right. Where he ends up with a decision that the parties can accept, albeit perhaps reluctantly, the dispute will go no further. Where, however, the adjudicator's decision does not lead to a resolution of the dispute, I would suggest that the proper place is arbitration. I am not talking about arbitration as practised by the dinosaurs of the 1970s and 1980s. I am talking about 1996 Act arbitration by arbitrators who have the courage and personality to bring even the most intransigent party to heel, and party representatives who accept that what the industry wants is dispute resolution processes that will avoid the descent into the trenches of attrition. I am talking about the rational human beings, not those who get so wrapped up in 'principles' that they forget that the building industry is about building buildings and not about screwing the last penny out of someone else. Those people are welcome to litigation but they may well find a shock in store with the 'Woolf' reforms and case management by judges.

This is another reason to stay with arbitration. The arbitrator will be someone who is steeped in the building industry; he can identify what it is that the parties are really arguing about and set procedures to suit. The judges in the High Court, however experienced, have only ever seen the problem jobs. They may be experts in the law but they will almost always have to have the technical content explained to them. If all construction disputes that are not solved by adjudication go to the courts, the waiting lists will go on forever unless a lot more judges are appointed. That is an unlikely scenario unless those using the courts pay for the privilege, thus undermining the only real advantage the courts have over arbitration. It is far better to appoint a technical arbitrator to sort things out.

Another innovation resulting from the passage of the Housing Grants, Construction and Regeneration Act is a statutory regime of payment provisions. Whilst innovative, these fit most comfortably with the other examples of documents relating to payment so these are all in Chapter 4.

At various points all contracts lay down procedures which must be followed, but things can go wrong. The contracts require that certain notices should be served in particular circumstances and there are many occasions where the contractor must, or should, set down the situation in writing. Sometimes this is spelled out in the contract but often it is not. Even where it is, it is all too often the case that the parties to the contract never read it. Although there is an old adage that the best jobs are those where the contract is signed and put in the drawer for the duration of the project, this can, where things start to go a bit awry, cause more problems than it solves. Current commercial pressures have, in any event, probably made that a thing of the past. Where the contract is silent, prudence and business sense dictate that certain things be recorded in writing there and then, even though the contract does not expressly require it.

As a generalisation, in a dispute situation the person with the best and most accurate records has a head start over his opponent. Memory is fallible, the written record endures. If something is stated in writing to the other party and is not challenged by him at the time, he is going to be in difficulty subsequently asserting that the written record is incorrect. It is certainly not impossible for him to do so but the burden on him is heavy.

A great many building contract disputes arise because one party or the other does not realise the fundamental truth that contracting is about the allocation of risks. The JCT contracts are a negotiated attempt to apportion risks fairly between the parties. The Joint Contracts Tribunal is representative of all users of the standard forms. The standard forms are not an imposed contract drawn up by one party in its own interests and imposed on the other. They are forms prepared and revised jointly by several representative bodies and accepted by all those involved: employers, contractors, architects and other professionals.

The chapters that follow endeavour to cover some of the situations that commonly arise. This is not an attempt at a legal textbook nor a complete procedural handbook. Indeed where the substance of a letter or notice required by the contract, or desirable in particular circumstances, is obvious, I follow the example of my predecessors carefully stated in the previous editions and 'have not bothered to give an example, as to do so would be an insult to the intelligence of our intended readers, as well as to waste valuable space'.

Instead, the text concentrates on those situations that give rise to common problems and in particular those in which things are likely to go wrong or where, because the situation is so obvious, it is often overlooked. A simple commentary is given followed by suggested possible wording for some notice or letter. In certain cases common forms are reproduced so that the reader can put them in the contractual framework.

As my predecessors were at pains to point out, do not follow the suggested wording slavishly. In some cases the examples are extreme. It is suggested that the reader should, so far as possible, follow the actual wording of the contract when giving notices and the like, even though the wording may, in some cases, be tortuous and ungrammatical.

All the parties to a construction contract are human; they make mistakes, do things at the wrong time or simply overlook them. This book attempts to retain an overall view, bearing in mind in particular that it is aimed primarily at contractors.

It is not the purpose of this book to discuss the merits or demerits of how the JCT contracts apportion risk. Its aim is to help the reader to work under the contracts as drafted and not to suggest how, perhaps, they should have been drafted.

1.02 Letters and notices

One point of particular note is how the various notices required by the contracts and other correspondence should be sent to the architect or employer. Common practice nowadays is to send correspondence that is of particular importance by fax and follow that up by first class post.

In many instances the JCT contracts require notices to be sent by 'actual delivery, special delivery or recorded delivery'. This must be done in the way required by the contract but it is always appropriate to send the notice by fax first. Actual delivery could involve delivering the letter yourself or using the services of a courier with a signed record of delivery. The Post Office offers a Special Delivery service which guarantees delivery by 12 noon the next working day after posting to most UK destinations. Recorded delivery is just what it says; the Post Office will tell you on enquiry that the letter has arrived. For an extra charge the Post Office will provide signed proof of delivery.

It is worth mentioning clause 1.7 of JCT 98 here. Clause 1.7 states that where the contract does not specify the manner of giving or service of any notice, service shall be by any effective means to any agreed address. Where no address has been agreed service shall have been effected if the notice is sent to the addressee's last known principal business address or in the case of a body corporate, to that body's registered or principal office. (The same provision is set out in clause 1.13 of IFC 98, in clause 1.5 of MW 98, in clause 1.5 of JCT 98 WCD, in clauses 1.4 in NSC/C and NAM/SC and in clause 24.1 of DOM/1.) Notices required by the contract should be given to the employer where this is specified but a copy should always be sent to the architect.

When notifying the architect or employer of any occurrence or giving a notice that is required by the contract, you should take care to read the requirements of the particular clause relating to what specifically needs to be done. The example documents in this book include in many instances a notation to the effect that they should be sent by Special Delivery. (You must read this to refer to actual or recorded delivery if that is more appropriate in specific circumstances.) This notation is not limited to those letters that are notices required by the contract to be sent in this way. Any of the letters that are given as examples in this book will, as noted above, almost invariably be sent by fax as well.

A further development in terms of communication is the internet and e-mail. I am sure that this will be used more and more frequently in years to come. There is nothing to prevent its use but given its relative unfamiliarity at the present time it would be advisable to obtain agreement before committing any communication to this method.

1.03   Tabular summary of mandatory requirements

There are many notices and letters that the contractor is obliged to give under the JCT standard forms of contract. Set out here, in tabular form, are the requirements of this nature that arise out of JCT 98. Where there are similar requirements that arise out of the other JCT standard forms, these are set out alongside.

In a further table in Chapter 8 are additional obligations concerning notices that are imposed on the contractor under the subcontracts that he enters into with his domestic and nominated subcontractors.

JCT 98 clause	Content	WCD 98 clause	IFC 98 clause	MW 98 clause
2.3	Notice specifying disrepancy or divergence in documents			
	Notice of divergence between Employer's Requirements and site boundary	2.3.2		
2.4.1	Notice specifying discrepancy or divergence between statement in respect of performance specified work and instruction issued by architect			
	Notice of proposed amendments to deal with discrepancy in Employer's Requirements	2.4.1		
2.4.2	Inform architect of correction to statement			
	Notice of proposed amendment to deal with discrepancy in Contractor's Proposals	2.4.2		
	Notice of discrepancies within Employer's Requirements or Contractor's Proposals	2.4.3		
4.1.1.1	Objection to compliance with instruction requiring change in obligations or restrictions	4.1.1	3.5.1	
4.2	Request for contractual authority for instructions	4.2	3.5.2	
4.3.2	Confirmation of oral instructions	4.3.2		
5.3.1.2	Provide master programme and revisions thereto			
	Provide drawings, specifications etc.	5.3		

JCT 98 clause	Content	WCD 98 clause	IFC 98 clause	MW 98 clause
5.4.2	Advise architect of need to receive further drawings, details or instructions		1.7.2	
	Supply as-built drawings	5.5		
5.6	Return all drawings, details and descriptive schedules to architect if so requested			
5.9	Supply drawings and information related to performance specified work			
6.1.1	Give notices required by Acts of Parliament etc.	6.1.1	5.1	5.1
6.1.2	Notice specifying divergence between statutory requirements and contract documents	6.1.2	5.2	5.1
6.1.4.2	Notice of emergency work		5.4.2	
6.1.6	Notice specifying divergence between statutory requirements and contractors statement concerning performance specified work			
	Give notice to Health and Safety Executive			1.4
6A.2	Notify amendments to health and safety plan	6A.3	5.7.2	5.7
6A.4	Provide information to planning supervisor	6A.5.2	5.7.4	5.9
	Prepare and deliver health and safety file (when acting as planning supervisor)	6A.5.1		
8.2.1	Vouchers proving materials comply with required standard	8.2		
	Notify objection to instruction (when acting as planning supervisor)	12.2.2		
13.2.3	Disagreement with the application of clause 13A to an instruction			

JCT 98 clause	Content	WCD 98 clause	IFC 98 clause	MW 98 clause
13.4.1.2 A1	Submit price statement	12.4.2 A1	3.7.1.2 A1	
13.4.1.2 A4.2	State whether or not amended price statement accepted	12.4.2 A4.2	3.7.1.2 A4.2	
	Agree price of variation			3.6
13.5.2	Daywork vouchers	12.5.4		
18.1	Consent to employer taking partial possession	17.1	2.11	
19.1.1	Request for consent to assignment	18.1.1	3.1	3.1
19.2.2	Request for consent to subletting	18.2.1	3.2	
	Request for consent to sublet design	18.2.3		
19.3.2.1	Consent/request for consent to additional names of subcontractors			
21.2.1	Production of insurance documents/policies/premium receipts	21.1.2	6.2.2	6.4
21.2.2	Deposit of insurance policy and receipts	21.2.2	6.2.4	
22A.2	Deposit of insurance policy and receipts	22A.2	6.3A.2	
22A.3.1	Production of insurance evidence/policy and receipts	22A.3.1	6.3A.3	
22A.4.1	Notice of loss or damage	22A.4.1	6.3A.4.1	
22B.3.1	Notice of loss or damage	22B.3.1	6.3B.3	
22C.4	Notice of loss or damage	22C.4	6.3C.2	
22C.4.3.1	Notice of determination	22C.4.3.1	6.3C.2	
22D.1	Quotation for liquidated damages insurance	22D.1	6.3D.1	
22D.1	Deposit of insurance policy	22D.1	6.3D.1	
23.3.2	Consent to partial use or occupation	23.3.2	2.1	
23.3.3	Notify amount of additional premium/provide premium receipt	23.3.3	2.1	
25.2.1.1	Notice of delay	25.2.1	2.3	2.2

JCT 98 clause	Content	WCD 98 clause	IFC 98 clause	MW 98 clause
25.2.1.2	Copy notice to nominated subcontractor			
25.2.2	Particulars and estimates	25.2.2	2.3	
25.2.3	Further notice, particulars and estimates	25.2.3		
26.1	Application for direct loss and/ or expense	26.1	4.11	
26.1.2	Information in support of application	26.1.2	4.11	
26.1.3	Details of loss and/or expense	26.1.2	4.11	
26.4.1	Notification of nominated subcontractor's application for direct loss and/or expense	26.1.2	4.11	
27.3.2	Notification of composition or arrangement	27.3.2	7.3.2	
	Provide copies of drawings, details etc.	27.6.1		
27.7.2	Application for statement regarding applicability of clauses 27.6.1 to 27.6.6 or for statement of account	27.7.2		
	Application for statement regarding applicability of clauses 7.6(b) to 7.6(g) or for statement of account		7.7.2	
28.2.1	Notice of default or defaults	28.2.1	7.9.1	7.3.1
28.2.2	Notice of specified suspension event or events	28.2.2	7.9.2	
28.2.3	Notice of determination	28.2.3	7.9.3	7.3.1
28.2.4	Notice of determination (repeated event)	28.2.4	7.9.4	
28.3.3	Notice of determination (insolvency of employer)	28.3.3	7.10.3	7.3.2
28.4.3	Prepare an account setting out total value of work executed etc. at time of determination by contractor	28.4.4	7.11.3	7.3.3

JCT 98 clause	Content	WCD 98 clause	IFC 98 clause	MW 98 clause
28A.1.1	Notice consequent upon suspension of works	28A.1.1	7.13.1	
	Provide copies of drawings, details etc.	28A.3		
28A.5	Production of documents	28A.6	7.18	
29.2	Consent to work by employer	29.2	3.11	
30.1.1.6	Provision of advance payment bond	30.1.1.2	4.2(b)	
30.1.2.2	Submission of application setting out amount of gross valuation	30.3.1	4.2(c)	
	Make application for interim payment	30.3.1		
30.1.4	Notice of intention to suspend performance	30.3.8	4.4A	4.8
30.3.1/30.3.2	Provision of proof of ownership	15.2.1	4.2.1.1	
30.3.1	Provision of bond	15.2/ 15.2.2	4.2.1.1/ 4.2.1.2	
30.3.5	Provision of proof of insurance	15.2.5	4.2.1.5	
30.5.3	Request for retention to be placed in a separate bank account	30.4.2.2		
30.6.1.1	Submission of documents necessary for adjustment of contract sum		4.5	4.5.1.1
	Submit final account and final statement	30.5.1		
31.3.1	Notice and evidence of status under tax deduction scheme	31.3.1	B.2.1	C2.1
31.4.1	Notice of change of status	31.4.1	B.3.2	C3.1
31.4.2	Notice regarding entitlement or otherwise to be paid without statutory deduction	31.4.2	B.3	C3.2
31.4.3	Notice of cancellation of tax certificate	31.4.3	B.3.3	C3.3
31.6.1	Statement of direct cost of materials used in the works	31.6.1	B5.1	C5.2

JCT 98 clause	Content	WCD 98 clause	IFC 98 clause	MW 98 clause
34.1.3	Notice of discovery of antiquities etc.	34.1.3		
35.5.1	Reasonable objection to proposed nominated subcontractor			
	Reasonable objection to proposed named subcontractor		3.3.2(c)	
35.7	Send completed NSC/A and NSC/T Part 3 to architect			
	Notify employer of date of subcontract with named subcontractor		3.3.1	
35.8	Notify failure to reach agreement with proposed subcontractor			
	Notify failure to reach agreement with named subcontractor		3.3.1	
35.13.2	Discharge payment to nominated subcontractor			
35.13.3	Production of reasonable proof of payment			
35.14.2	Request for written consent to extension of time for nominated subcontractor			
35.15.1	Notice of nominated subcontractor's failure to complete			
35.18.1.2	Agreement or otherwise to substituted subcontractor's price			
35.24.1	Notice of nominated subcontractor's default			
	Advise of events likely to lead to determination of named subcontractor's employment		3.3.3	
35.24.2	Notice of nominated sub-contractor's insolvency			

JCT 98 clause	Content	WCD 98 clause	IFC 98 clause	MW 98 clause
35.24.6.2	Notice informing architect whether or not nominated subcontractor's employment determined			
	Notice informing architect that named subcontractor's employment determined		3.3.3	
	Account for amounts recovered from named subcontractor in respect of determination		3.3.6	
38.4.1/39.5.1	Notice of increases or decreases	36.4.1/ 37.5.1		A4.1
38.4.5/39.5.5	Provide evidence and computations	36.4.5/ 37.5.5		
	Provide certificate			A4.4.1
40.2.1	Provide amplification of contract sum analysis	38.2		
	Include statement of allocation of values and amount of adjustment	38.2		
40.3	Insert market price of articles manufactured outside UK	38.3		
	Agree alteration to methods and procedures	38.4		
41A.2.2	Execute adjudication agreement	39A.2.2	9A.2.2	D.2.2
41A.4.1	Notice of intention to refer dispute to adjudication	39A.4.1	9A.4.1	D4.1
41A.2.2	Agree adjudicator/apply to nominating body	39A.2.2	9A.2.2	D2.2
41B.1.1	Notice of arbitration	39B.1.1	9.1	E2.1
42.2	Provide contractor's statement			
42.5	Amend contractor's statement			
42.13	Provide analysis of portion of contract sum relating to performance specified work			
42.15	Notice of injurious affection			

JCT 98 clause	Content	WCD 98 clause	IFC 98 clause	MW 98 clause
	Where a WCD contract includes supplementary provisions the following apply			
	Submission of drawings, details, etc.	S2.1.1		
	Notify name of site manager	S3.2		
	Site manager to make records available	S3.4		
	Notify date of subcontract with named subcontractor	S4.2.1		
	Inform reasons for not entering into subcontract	S4.2.2		
	Notify name of alternative subcontractor	S4.3.1		
	Obtain consent for determination of employment of named subcontractor	S4.4.1		
	Account for amounts recovered from named subcontractor in respect of determination	S4.4.3		
	Submission of estimates	S6.2/S7.2/ S7.3		
	Raise objection	S6.2.2		

Notices required by any VAT agreement are not included in the above table.

Chapter 2
Pre-contract Documentation

2.01 Introduction

Competitive tendering by contractors on a selected list is traditional in the UK. The NJCC (the National Joint Consultative Committee for Building) in collaboration with equivalent bodies in Scotland and Northern Ireland, produced four Codes of Procedure for Selective Tendering. They were:

- Code of Procedure for Single Stage Selective Tendering (1996)
- Code of Procedure for Two Stage Selective Tendering (1996)
- Code of Procedure for Selective Tendering for Design and Build (1995)
- Code of Procedure for the Selection of a Management Contractor and Works Contractors (1991)

The NJCC ceased trading in 1996 but the four codes are, at the time of writing in 1999, still selling well and RIBA Publications will keep them in print as long as demand lasts. The successor to the NJCC is the Construction Industry Board (CIB). The CIB has published a total of 14 reports at the time of writing, the two of particular relevance to this book being the Code of Practice for the Selection of Main Contractors and the Code of Practice for Selection of Subcontractors. All 14 of the reports are available at the bookshops run by the various institutions connected with the construction industry and elsewhere. The Code of Practice for the Selection of Main Contractors, mentioned above, deals with a number of aspects of the tendering process in the following sections:

- Qualification and compilation of the tender list
- Tender invitation and submission
- Tender assessment

It contains an annex which sets out the minimum that should be provided to tenderers by way of documents and information.

The CIB is based at 26 Store Street, London WC1E 7BT.

Much of what follows in this chapter assumes that the procedures laid down in the code have been used.

2.02 Qualification

You will sometimes find that you are asked whether you wish to be included on a select list of contractors, whether for one particular contract or for future projects.

If the recommendations of the CIB Code of Practice are followed by the employer you should be judged on certain criteria when being considered for inclusion on a tender list. ˜

The Code suggests that the criteria for qualification should include:

- Quality of work
- For design and construct procurement, the capability to offer design
- Performance record
- Overall competence
- Health and safety record and competence
- Financial stability
- Appropriate insurance cover
- Size and resources
- Technical and organisational ability
- Ability to innovate

2.03 Preliminary enquiry

The CIB Code of Practice also sets out criteria specific to the project against which prospective tenderers should be tested. Those identified by the CIB are:

- Their current and expected capacity
- The relevance of their skills and experience to the demands of the proposed project
- The team that they would make available
- Their grasp of, and enthusiasm for, the proposed project

If the CIB code is followed, you will receive a preliminary enquiry that sets out a questionnaire including these points. This enquiry will also set out such things as the job name and location, the nature, scope and location of the project, programme details and the like.

The NJCC Code of Procedure for Single Stage Selective Tendering which, as noted above, is still in common use, sets out a draft enquiry letter for use in preparing a preliminary list of tenderers (**document 2.03.1**). The other NJCC Codes of Procedure have similar letters. One thing that this draft does not include is any requirement for the production of project specific criteria and you must expect such enquiries to be made.

2.03.1 Form of tender (this form of tender is suitable for use only when a formal contract is entered into

Appendix A.1 PRELIMINARY ENQUIRY FOR INVITATION TO TENDER
Applicable where the JCT Standard Form of Building Contract, Intermediate Form of Building Contract or Agreement for Minor Building Works is to be used

Dear Sirs,

Heading

I am/We are authorised to prepare a preliminary list of tenderers for construction of the works described below.

Please indicate whether or not you wish to be invited to submit a tender for these works. Your acceptance will imply your agreement to submit a wholly bona fide tender in accordance with the principles laid down in the NJCC 'Code of Procedure for Single Stage Selective Tendering' 1996, and not to divulge your tender price to any person or body before the time for submission of tenders. Once the contract has been let, I/we undertake to supply all tenderers with a list of the tender prices.

You are requested to reply by . . . Your inability to accept will in no way prejudice your opportunities for tendering for further work under my/our direction; neither will your inclusion in the preliminary list at this stage guarantee that you will subsequently receive a formal invitation to tender for these works.

Yours faithfully . . .

a Project . . .
b Type and function of the building, e.g. commercial, industrial, housing, etc., with any other details . . .
c General description of the project . . .
d Employer . . .
e Employer's professional team . . .
f Planning Supervisor . . .
g Location of site . . . (Site plan enclosed)
h Approximate cost range of project £ . . . to £ . . .
i Number of tenderers it is proposed to invite . . .
j Nominated sub-contractors for major items . . . [1]
k Form of Contract:
List here Form of Contract, e.g. JCT 80/IFC 84/MW 80,* together with
(i) Appendix items to the appropriate form completed †
(not applicable to MW 80)
(ii) The JCT Amendments proposed to the standard form ‡
[Note: Attention is drawn to the advice given in clause 4.2.3.]
l Examination and correction of priced bill(s) [Section 6 of Code] [2]; Alternative 1/Alternative 2 [3] will apply
m The contract is to be executed as a deed/a simple contract §
n Anticipated date for possession . . .
o Period for completion of works . . .
p Approximate date for despatch of all tender documents . . .
q Tender period . . . weeks
r Tender to remain open for . . . weeks [4]
s Details of guarantee requirements . . . (if any) [5]
t Particular conditions applying to the contract are . . .

The draft enquiry letter (**document 2.03.1**) refers to JCT 80 etc., which will obviously be replaced by JCT 98 etc. in the letter that you receive.

Although there appear at the time of writing to be no plans by the CIB to produce their own version of the NJCC Codes it is quite likely that this may be done in due course and you may find if this should happen, that a different format of letter is sent to you. It will obviously still contain much the same information but will no doubt contain an enquiry relating to project specific information.

If replying in the affirmative confirming you interest in being included on the tender list, **document 2.03.2** is a sample of an appropriate letter. **Document 2.03.3** is a possible letter of tactful refusal.

2.03.2   Letter of willingness to tender

To the Architect   Date

Dear Sir,

[Repeat heading on letter of enquiry]

Thank you for your letter of *[date]* asking us whether we wish to be invited to submit a tender for this project in accordance with the conditions which you set out.

We shall be pleased to accept such an invitation if extended and we look forward to hearing from you further.

As far as your request for some detailed information of this company is concerned we enclose our firm's brochure together with a list of similar projects that we have completed and an organisational chart showing the directors and staff. We confirm that we have the capacity to carry out the works in accordance with your outline programme and, provided that the project starts to your programme, we enclose an indication of the individuals who will be involved in managing the works.

Yours faithfully,

2.03.3   Letter of refusal

To the Architect   *Date*

Dear Sir,

[*Heading*

[*First paragraph as 2.03.2*]

We very much regret that our commitments at the present time
are such that we are reluctantly compelled to decline.
However, we hope that we may be given the opportunity to tender
for any future projects with which you may be concerned.

Yours faithfully,

2.04   Formal invitation to tender

The next stage is the receipt by you of a formal invitation to tender which will be accompanied by the tender documents.

Annex 1 to the CIB Code of Practice sets out recommended minimum requirements for documents and information for tender enquiry purposes (see **document 2.04.1**).

You may find that the enquirer uses the draft covering letter set out in the NJCC Code of Procedure (see **Document 2.04.2**).

The main points to note are:

- You should make arrangements to inspect the drawings and details at the specified place, which will usually be the architect's office. Although the documents sent to you should contain all the essential information, you have only yourself to blame if the drawings and details contain relevant information which may well affect your tender price and cause problems financially or otherwise during the course of the project.
- The site should be inspected. Subject to those matters which SMM 7 requires to be stated in bills of quantities, assuming that method of measurement is used, site conditions in general are at your own risk.
- You should be very careful to note which of the alternatives for the correction of errors in the Code of Procedure, if that is used, is made applicable. Alternative 1 means that you will only be given the opportunity of confirming or withdrawing your tender in the event of errors which affect your price. Alternative 2 allows for the correction of errors by adjustment of the tender price.

Document 2.04.3 is a suggested letter of acknowledgement.

2.04.01   Annex 1.   Tender enquiry documents

Tender enquiries should include the following documents and information as a minimum:

1. A full list of tender enquiry documents so that the tenderer can check that they are complete.

2. Instructions to tenderers, to include:
- where and by when to submit the tender
- the type of tender required, e.g. lump sum, priced bills, schedule of rates (and, if more than one of these is required, which will take precedence)
- all information required from tenderers (e.g. method statement, programme of work, identity of subcontractors)
- how the tender should be packaged and identified
- how any queries on the tender enquiry documents will be handled
- how any errors or inconsistencies in the tender enquiry documents, discovered after they have been issued, will be dealt with
- whether alternative proposals are acceptable together with a compliant tender
- the period of validity required for the tender
- the target date for tender acceptance
- details of arrangements for visiting the site during the tender period
- whether the client will accept the lowest or any tender
- the deadline after which requests for additional information will not be considered

3. Information for tenderers, to include:

- the name of the person to provide the single point of contact from within the client organisation for all communication
- latest information available about the scope of work
- proposed dates and durations of the works.
- any requirement to have a preferred tenderer approved
- the contract conditions to be used
- if amendments to the standard form or suite of forms are proposed, they should be specifically identified
- dispute resolution procedures, if not incorporated in the contract
- whether price fluctuations will apply and, if so, how
- payments terms, including information about retentions, retention bonds, advance payments for materials, security of payment and protection against non-payment
- any requirements for bonds, guarantees, insurances and the like
- terms and conditions of supply of any pre-ordered equipment and materials
- tender assessment criteria
- how any tenderer's errors will be handled
- how tender results will be communicated.

4. Standards and specifications.

5. Schedules and drawings, including any pricing schedules required by the form of contract (e.g. bills of quantities, although clients are invited to be selective in their requirements as such detailed submissions add to the 'hidden' costs of tendering). Where drawings are not included with the bills of quantities, it should be stated where the drawings on which the bills are based may be inspected.

6. The health and safety plan, a requirement of the *Construction (Design and Management) Regulations 1994*, including the identity of the planning supervisor, if known, and confirmation that the main contractor will be the principal contractor under the regulations (or other arrangements, if different).

7. Form of tender

2.04.02   Appendix B   FORMAL INVITATION TO TENDER

Dear Sirs,

Heading

Following your acceptance of the invitation to tender for the above, I/we now have pleasure in enclosing the following:

a   two copies of the bill(s) of quantities/specifications/schedules [1];

b   (i)   two copies of the location drawings, component drawings, dimension drawings and information schedules (e.g. JCT 80 With Quantities ‡ or IFC 84 where Contract Bills are provided) [1];

          (ii)   two copies of all drawings (for JCT 80 Without Quantities, IFC 84 with Specification and Drawings or MW 80) § [1];

          (iii) two copies of the planning supervisor's pre-tender health and safety plan;

c   two copies of the form of tender;

d   addressed envelope(s) for the return of the tender (*and priced bill(s) †*) and instructions relating thereto.

Will you please also note:

1   drawings and details may be inspected at . . .

2   the site may be inspected by arrangement with the employer/architect [1];

3   tendering procedure will be in accordance with the principles of the NJCC 'Code of Procedure for Single Stage Selective Tendering' 1996;

4   examination and adjustment of priced bill(s) [Section 6 of Code]; Alternative 1/Alternative 2 [1] will apply.

The completed form of tender is to be sealed in the endorsed envelope provided and delivered or sent by post to reach . . . not later than . . . hours on . . . the . . . day of . . . [date] . . .

*   *The completed form of tender and the priced bill(s) of quantities sealed in separate endorsed envelopes provided are to be lodged not later than . . . hours on . . . the . . . day of . . . 19 . . . The envelope containing the tender should be endorsed with the job title; that containing the bill(s) of quantities should be endorsed with the job title and tenderer's name.*

Will you please acknowledge receipt of this letter and enclosures and confirm that you are able to submit a tender in accordance with these instructions.

Yours faithfully.

Architect/Quantity Surveyor [1]

2.04.3   Letter of acknowledgement of invitation to tender

To the Architect *Date*

Dear Sir,

[Heading]

We acknowledge receipt of your formal invitation to tender of [*date*] with enclosures as listed therein.

We confirm that our tender will be submitted to you by the specified date in accordance with your instructions.

[If appropriate confirm any arrangements made for inspecting drawings etc. and the site.]

Yours faithfully,

2.05   Errors in tender documents

Architects and quantity surveyors are not infallible. For example you may discover errors in the documents enclosed with the invitation to tender. The documents may not comply with the minimum requirements in the CIB Code of Practice. Your site inspection may also reveal matters which ought to have been mentioned in the documents. If so, you should notify the architect immediately and, if the error is in the bills, the quantity surveyor also. This should, at least, be done where the undetected error could substantially affect the tender price. You may also need clarification of certain points. **Document 2.05.1** may be adapted as necessary.

2.05.1   Queries arising from tender documents

To the Architect/Quantity Surveyor   *Date*

Dear Sir,

[*Heading*

On examination of the tender documents enclosed with your
letter of [*date*] [*or*] Following inspection of the site], we
wish to draw the following matters to your attention:
[*Specify queries, etc.*]

[*Alternative:*
On examination of the Bills enclosed with your letter of [*date*]
we find the following description is ambiguous: [*Give
description of item by reference to item and page of bill,
etc.*]

We shall be grateful if you can clarify this item so that we may
be certain that we are tendering on the correct basis.

Yours faithfully,

2.06   Time for tendering

The CIB code (paragraph 3.5) sets out a schedule of suitable periods for tender preparation as follows:

Contract type	Preliminary enquiry Time to return preliminary enquiry	Tender Time to return tender
Design and construct	min. 3 weeks	min. 12 weeks
Construct only	min. 3 weeks	min. 8 weeks

You may be given less time. This may be acceptable in the case of smaller or less complex projects but you should always be aware of the problems that can arise later when time for proper preparation of a tender is not allowed.

If less time is given and you are unable to put together a proper tender as a result you should never be afraid to seek an extension of time since the other prospective tenderers are probably experiencing the same difficulties. Any such application should always be made at the earliest possible time. The phrasing of your letter will depend on the particular circumstances, but a typical letter is given as **document 2.06.1**.

2.06.1 Letter where the time for tendering is too short

To the Architect *Date*

Dear Sir,

[*Heading*]

We refer to your letter of [*date*] regarding the above project, and specifying that tenders must be submitted by [*date and time*]

Unfortunately, we find that it is impossible to obtain tenders from subcontractors for major elements of the work by the date specified for submission of tenders. The work in question is: [*specify*]. We very much regret, therefore, that we have no alternative but to ask you to postpone the date for submission of our tender until [*specify date and time*] as otherwise we shall be unable to submit a tender for this project.

We shall be grateful for an early response to this request.

Yours faithfully,

2.07   Onerous conditions in tender documents

One matter to watch out for is the attempt to apply onerous conditions. Whether a clause is onerous or not often depends on the perspective of the person reading it.

Whatever happens it is vital that you identify at the earliest possible stage any clause that in your opinion comes within this definition. If a clause is so onerous that you cannot accept it, the earlier that you notify the architect the better. He may be prepared to withdraw the offending requirement. If he is not, the earlier that you find out the more likely you are to avoid abortive costs if you decide not to submit a tender. If the clause is something that you can live with you will need to ensure that you make the appropriate allowance in your tender to cover any additional costs that you will incur as a result of the requirement.

Document 2.07.1 is a suggested letter to cover this situation.

2.07.1   Letter to architect concerning onerous conditions

To the Architect   *Date*

Dear Sir,

[*Heading*]

We are concerned to note that you [propose to levy liquidated and ascertained damages at the rate of £10,000 per week]. We are very concerned that this figure appears very high for a project the estimated value of which in your own invitation to tender is £40,000.

We would ask you to reconsider this figure as we shall be unable to enter into the contract if it is not altered.

We look forward to your early response.

Yours faithfully,

2.08   Submission of tender

The CIB code states that standard forms of tender can assist in ensuring that information from tenderers is comparable and that where possible such forms should be used. Many architects and quantity surveyors will continue to use a form of tender based on Appendix C of the NJCC Code of Procedure. You should complete it without any alterations or deletions other than those indicated; your tender must not be qualified in any way because, if it is, the NJCC Code prescribes that it must be rejected.

An unequivocal acceptance by the employer of a tender in this form creates a contract on the terms specified (except in Scotland) although the procedure envisages that a standard form of contract will be entered into. If for some reason it is not you will nonetheless be in a contractual relationship with the employer.

The standard form of tender is set out as **document 2.08.1.**

2.08.01 Appendix C FORM OF TENDER

This form of tender is suitable for use only when a formal contract is entered into

Tender for . . . (description of Works)

To . . . (Employer)

Sir(s),

I/We having read the conditions of contract and bill(s) of quantities/specifications/schedules [1] delivered to me/us and having examined the drawings and other documents referred to therein do hereby offer to execute and complete in accordance with the conditions of contract the whole of the works described for the sum of . . . £ . . . and within . . . weeks [2] from the date of possession.

*   *I/We undertake in the event of your acceptance to execute with you a formal contract embodying all the conditions and terms contained in this offer.*

I/We agree that should obvious errors in pricing or errors in arithmetic be discovered before acceptance of this offer in the priced bill(s) of quantities submitted by me/us these errors will be dealt with in accordance with Alternative 1/Alternative 2 [1] contained in Section 6 of the NJCC 'Code of Procedure for Single Stage Selective Tendering' 1996.

This tender remains open for consideration for . . . days [3] from the date fixed for the submission or lodgement of tenders [4].

Dated this . . . . day of . . . . 19 . . . . . .

Name . . . .

Address . . . .

Signature . . . . *Witness** . . . .

*   *The completed form of tender and the priced bill(s) of quantities sealed in separate endorsed envelopes provided are to be lodged not later than . . . hours on . . . the . . . day of . . . [date] . . . The envelope containing the tender should be endorsed with the job title: that containing the bill(s) of quantities should be endorsed with the job title and tenderer's name.*

2.09   Subcontractors and suppliers

In preparing your tender, you will no doubt be seeking tenders yourself from domestic subcontractors and suppliers. It is essential that they also observe the principles of the codes and that their tenders are submitted on conditions compatible with those of the eventual main contract. The CIB Code of Practice for the Selection of Subcontractors sets out good practice in this respect. This Code also contains recommended periods for tendering and it may be necessary for an extension of tendering time to be sought from the employer if subcontractors are to have an appropriate time to tender themselves, especially in a design and construct situation.

You must be very careful to ensure that, when inviting tenders from potential subcontractors and suppliers, they are made fully aware of all the conditions which you yourself will have to meet under the main contract, particularly any restrictions on your methods of working, access to the site and programming.

It is common for both domestic subcontractors and suppliers to ignore the conditions which you have specified and to submit their tenders on their own form. You can either exclude that tender from consideration or ask them to re-submit on the specified conditions.

Always ensure that the tenders submitted by domestic subcontractors and suppliers are open for acceptance for a long enough period to permit you to accept them after you have yourself been notified that your tender has been accepted.

2.10   Letters of intent

It is quite usual for you to receive a letter from the architect notifying you of the employer's intention to enter into a contract with you on the basis of your tender. Sometimes the letter of intent will ask you to undertake certain preliminary work.

Letters of intent should be treated with great caution since they rarely contain any express undertaking by the employer to pay for any work done in anticipation of the contract. If the contract does not proceed, you may find it difficult to obtain payment. A worst case scenario is that you could be involved in arbitration or court proceedings.

A suggested letter of acknowledgement which may help to safeguard your interests is given as **document 2.10.1**.

A modified form of this letter can be used if you are asked to start work on site before the contract documents are signed.

2.10.1 Acknowledgement of letter of intent

To the Architect *Date*

Dear Sir,

[*Insert appropriate heading*]

Thank you for your letter of *date*] informing us of the
Employer's intention to enter into a contract with us for this
project on the basis of our tender of [*date*].

You request us to [*insert nature of work requested*] in
anticipation of the contract. However, before undertaking
this preparatory work we would like to have an express
undertaking from the Employer to reimburse the cost of any work
so undertaken, together with a reasonable allowance for
overheads and profit, should the contract not proceed for any
reason.

If we do proceed in this way we are concerned to make you aware
that the cost of preliminaries items, if our work is for any
reason limited to the work requested by you, may not be
directly related to the figures inserted in the Bills of
quantities.

On receipt of this undertaking, we shall proceed immediately
in accordance with your request.

Yours faithfully,

2.11   Signing the contract

Completion of the contract documents is the employer's responsibility. When dealing with public authorities, the contract documents will be prepared by the employer's legal department. Nowadays it is almost always the case in the private sector that the quantity surveyor, where one is engaged, prepares the contract documents.

You should check the contract documentation before signing. In particular, you should check the deletions and entries in the appendix. The information to be stated there should have been set out in the bills of quantities on which you tendered.

You should also check that any amendments to the printed conditions have been made as notified to you at the time of tender, and that no amendments have been made of which you have not been notified. If this information is not the same as was indicated in the tender documents you have a valid reason to complain. Where there have been negotiations on price and especially if the terms of the contract have been subject to negotiation since the tender was submitted, it is vital to check that the contract document that you sign is precisely that which you expect.

Any deletions and amendments must be initialled by you and the employer (*not* the architect).

Document 2.11.1 is a letter typical where the contract documents are not consistent with the information given for tendering.

It is bad practice for the contract to be executed after the work has begun, but it often happens. If you are asked to commence work before the contract is signed or executed as a deed (sealed), you should check that there is a binding contract in existence, i.e. that you and the employer have agreed on the minimum essential terms and that your tender has been unequivocally accepted by the employer. You would be wise to protect your position, and **document 2.11.2** is the sort of letter you should send dependent on the circumstances. Once the contract is signed or executed after work has commenced, its terms will act retrospectively unless – unusually – a special condition provides otherwise.

Often the contract is merely signed by the parties, but sometimes it will be executed. The only practical difference between a signed and a contract which is executed as a deed is that the limitation period (i.e. the time during which an action can be brought against you or by you against the employer for breach of the contract) is 6 years for a signed contract and 12 years for a deed.

2.11.1 Letter where contract documents are inconsistent with tendering information

---

To the Architect *Date*

Dear Sir,

[*Heading*]

We acknowledge receipt of your letter of [*date*] enclosing the contract documents for signature [*or* executing as a deed] by us.

However, we must point out that [*state nature of inconsistency as appropriate, e.g.* Clause 25.4 (JCT 98) has been substantially amended in a manner of which we were not notified at the time of tender and which substantially reduces our entitlement to extensions of time].

We regret that we are not prepared to enter into a contract on this basis without substantial adjustment of the Contract Sum and we therefore return the documents to you for reconsideration by the Employer. We are, of course, prepared to contract if the amendment in question is removed.

Yours faithfully,

2.11.2 Letter where asked to start work before the contract is executed

---

To the Architect *Date*

Dear Sir,

[*Heading*]

Thank you for your letter of [*date*] requesting us to start work on site before the contract documents are completed.

[We are advised that there is already a binding contract between ourselves and the employer incorporating [The Standard Form of Building Contract, 1998 edition, private edition with quantities [amended as discussed and agreed between ourselves as recorded in the attached copy correspondence] on the basis of our tender dated [*insert date*] and the employer's formal acceptance dated [*insert date*] and we ask that the employer should formally confirm to us that this is his understanding of the contractual position so that we may commence work on [*date*] as requested.]

[*Or:*
Since the date of commencement is still some six weeks away we see no reason why the contractual formalities should not be completed within the course of the next few days so that we may start work on the date you request.]

[*Or:*
We would point out, however, that there is as yet no concluded contract between the employer and ourselves because [*give reason, e.g.* there are still the matters raised in our letter of [*date*] that have not been resolved. Hopefully we will be able to agree this outstanding matter with the employer and complete the contractual formalities in sufficient time to enable us to commence work as requested.]

Yours faithfully,

In JCT 98, IFC 98 and WCD 98 a single page is provided which allows you and the employer either to sign or to execute the contract as a deed (in MW 98 there is no provision for executing). If you do not have a seal – the contract can still be executed as a deed by the appropriate company officers.

You should have been notified when tendering whether the contract was to be signed or executed. If you do find that you are required to contract by a deed when you have not been so notified, it is as well to draw the architect's attention to this and endeavour to contract under signature only, but if the employer is insistent it is a matter for your own commercial judgement.

2.12   Performance bonds

Sometimes the employer will require you to provide a bond for due performance of the contract and this will have been notified to you at the time of tender.

Bonds can be obtained from both banks and certain insurance companies specialising in this type of business. A cross- or counter-guarantee may also be required from a parent company in respect of its subsidiary.

A typical bond requirement in the bills or other contract documents might read:

> 'Provide for an approved bond or bank security in a sum amounting to 10% of the Contract Sum, for the due execution and completion of the works and for the payment of any damages, losses, costs, charges or expenses for which the contractors may become liable under the contract. The bond will not be released until a Certificate of Making Good Defects has been issued for the whole of the works.'

Documents 2.12.1 and **2.12.2** illustrate bonds. **Document 2.12.1** is a typical local authority form of bond, but you should appreciate that the exact wording of both it and any supporting documents will vary according to their source.

Bonds must be executed as deeds because they contain a promise by the bank to the employer for which no reciprocal promise or benefit has been given.

2.12.1   Form of bond

BY THIS BOND We
whose registered office is at

(hereinafter called 'the Contractor')

and
whose office in the United Kingdom is at

(hereinafter called 'the Surety') are

held and firmly bound unto

(hereinafter called 'the Employer')

in the sum of

for the payment of which sum the Contractor and Surety bind themselves their successors and assigns jointly and severally by these presents.

Sealed with our respective seals and dated this
day of   [date]

WHEREAS the Contractor by an Agreement dated
and made between the Employer of the one part and the Contractor of the other part has entered into a Contract (hereinafter called 'the said Contract') for the construction completion and maintenance of certain Works as therein mentioned in conformity with the provisions of the said Contract.

NOW THE CONDITION of the above-written Bond is such that if the Contractor shall duly perform and observe all the terms provisions conditions and stipulations of the said Contract on the Contractor's part to be performed and observed according to the true purport intent and meaning thereof or if on default by the Contractor the Surety shall satisfy and discharge the damages sustained by the Employer thereby up to the amount of the above-written Bond then this obligation shall be null and void but otherwise shall be and remain in full force and effect but no alteration in terms of the said Contract made by agreement between the Employer and the Contractor or in the extent or nature of the Works to be constructed completed and maintained thereunder and no allowance of time by the Employer or the Architect under the said Contract nor any forbearance or forgiveness in or in respect of any matter or thing concerning the said Contract on the part of the Employer or the said Architect shall in any way release the Surety from any liability under the above-written Bond.

The Common Seal of
[the Contractor]
was hereunto affixed
in the presence of:

The Common Seal of
[the Surety]
was hereunto affixed
in the presence of:

2.12.2   Another form of bond

THIS DEED is made the . . . . date
of . . . . [date] . . . . BETWEEN . . . . whose
registered office is at. . . . . (hereinafter
called 'the Guarantor') of the one part and . . . .
. . . . whose registered office is at
. . . .
(hereinafter called 'the Employer') of the other part

   WHEREAS

(1) This Agreement is supplemental to a contract (hereinafter
called 'the Contract') dated the . . . . day of . . . . [date] . . .
and made between the Employer of the one part and . . . .
whose registered office is at . . . .
(hereinafter called 'the Contractors') of the other part whereby the Contractors agreed
and undertook to carry out the following works . . . .
. . . .
. . . .
(2) The Guarantor has agreed the due performance of the contract in manner
hereinafter appearing

NOW THIS DEED WITNESSETH as follows:-

1. The Guarantor hereby covenants with the Employer that the Contractors will duly
perform the obligations on the part of the Contractors contained in the Contract and that
if the Contractors shall in any respect fail to execute the Contract or commit any breach
of any of their obligations thereunder then the Guarantor will be responsible for and will
indemnify and keep indemnified the Employer from and against all losses damages
costs and expenses which may be suffered or incurred by it by reason or arising directly
or indirectly out of any default on the part of the Contractors in performing and observing
the obligations on their part contained in the Contract.

In WITNESS whereof the Guarantor
have caused their Common Seal to
be hereunto affixed this day and
year first above written

THE COMMON SEAL OF
the Contractors
was hereunto affixed
in the presence of:

2.13 Advance payment bonds and bonds in respect of payment for off-site materials and/or goods

In consequence of the Latham Report *Constructing the Team* the JCT has introduced a provision for the advanced payment of contractors in respect of work before it is done. Clause 30.1.1.6 of JCT 98 (clause 30.1.1.2 of WCD 98 and clause 4.2.1(b) of IFC 98) covers this and provides for the giving of a bond by the contractor in return. The form of those bonds is precisely the same in all these contracts and is set out in the contract form. I would have liked to reproduce it here but the payment that the JCT now requires for permission to do so is quite out of proportion.

The introduction of the 1998 suite of JCT contracts also marks a considerable change from the previous regime in the way of dealing with the payment for materials before they are delivered to site. There is no longer any opportunity to have materials paid for before delivery to site unless the employer decides that this can be done. The only materials and goods that can be paid for before delivery to site are those that have been prefabricated for inclusion in the works and listed by the employer in a list supplied to the contractor and annexed to the Contract Bills, the Employer's Requirements or other such contract document. These material have generally to be 'uniquely identified' and the contract gives the employer the opportunity to require a bond securing the value of such materials or goods which may be in the form set out in the contract or in an alternative form to the specific requirements of the employer. If the materials or goods are not uniquely identified the bond *has* to be provided before any entitlement to payment accrues. Again I should have liked to reproduce the standard form of these bonds but the JCT did not make this economically feasible.

Chapter 3
Commencement and Progress

3.01 Pre-commencement meeting

As mentioned in Chapter 2 it is not unusual for work to commence on site before the formal contract is signed. Between acceptance of the tender and work starting on site it is usual for the architect to convene a meeting principally for the purpose of getting to know the contractor's project team but also to review the general state of progress of such matters as signing the contract and the detailed design.

This latter point is the subject of a new departure introduced in Amendment 18 to JCT 80 and now firmly ensconced in the sixth recital and clause 5.4.2 of JCT 98. It formalises the not uncommon situation that the detailed design of those elements of the building that will not be started until the later stages of the project is continuing while the structure of the building is in course of construction. It is an optional provision that puts the drip feed of information from the architect onto a contractual footing by means of what is known as an 'information release schedule'. Any delays in the provision of late information can, if this optional provision is included in the contract, be more easily monitored than has been the case in the past. At the very least it avoids those endless arguments about whether information was provided on time or not.

It is of course vital that your programme and the information release schedule fit together.

A typical agenda for a pre-commencement meeting is shown as **document 3.01.1**.

3.01.1   Specimen agenda for pre-commencement meeting

AGENDA FOR PRE-COMMENCEMENT MEETING
To be held on [*date*]
At [*location*]

Job Aqua Products New Factory
Job No 456

---

1.   Introduction of professional team

2.   Introduction of contractor's management and site team

3.   Formal contract documents

4.   Master programme and other programming items

5.   Information release schedule

6.   Health and safety

7.   Subcontractors
        Nominated
        Domestic   Clause 19.2.1
             Named   Clause 19.3.1

8.   Dates for site meetings

9.   Dates for valuations/certificates
        Procedure (see clause 30.1.2.2)

10.   Any other business

Distribution:

Employer
QS
Consultants
Contractor
Clerk of Works
Subcontractors and suppliers
Architect's file

3.02   Site and progress meetings

It is customary for regular meetings to be held on site at which problems can be discussed and, hopefully, resolved by mutual agreement. They are usually chaired by the architect who will often prepare the record of the meeting. Some architects prefer this to be done by the contractor.

A typical agenda for a site meeting is shown as **document 3.02.1**. This agenda is based on that in the 7th edition 1996 of *Contract Administration* by the Aqua Group. I have included the new items 'Information Release Schedule' and 'Health and Safety Matters' which I suggest are items that should be included on such an agenda in line with the 1998 JCT Forms. Ideally, the agenda should be circulated about a week before the date of the meeting, having been agreed between you and the architect.

The record of the meeting is commonly referred to as 'the minutes'. It is important that the record should be accurate and, if prepared by the architect, you should examine it carefully and challenge any inaccuracies in writing immediately (**document 3.02.2**). Do not wait until the next meeting.

If inaccuracies are allowed to go unchallenged, and the record needs to be referred to subsequently, e.g. in a claims situation, it may be difficult if not impossible to prove that the record is in fact inaccurate. The record should be impartial. **Document 3.02.3** is a typical example.

3.02.1   Specimen agenda for site meeting

Project: Shops & Offices,   Newbridge St,   Borchester
Project ref: 456
Agenda for site meeting
Date: 7 October 1998 at 10.00 am

1.0	Apologies
2.0	Agree minutes of last meeting

3.0	Contractor's report:
	   General report
	   Sub-contractors' meeting report
	   Progress and causes of delays and claims arising
	   Information received since last meeting
	   Information and drawings required
	   Architect's Instructions required

4.0	Clerk of Works' report
	   Site matters
	   Quality control
	   Lost time

5.0	Consultant's reports
	   Architect
	     Information release schedule
	   Structural Engineer
	   Services Engineer

6.0	Quantity Surveyor's report
7.0	Health and safety matters
8.0	Communications and procedure

9.0	Contract completion date
	   Assess likely delays on contract
	   Review factors from previous meeting
	   List factors for review at next meeting
	   Record completion date (as revised)

10.0	Any other business
11.0	Date, time and place of next site meeting Date & time of next official visit

Distribution

Client
Main Contractor
QS
Struct Engr
Services Engr
Clerk of Works
File

3.02.2   Letter to the architect challenging the accuracy of the record

---

To the Architect *Date*

Dear Sir,

[*Heading*]

Thank you for the record of the site meeting held on [*date*].

We have the following comments to make:

[Item 3.04 <u>Cladding Work to H Block</u>
Mr Wood did not state that this was four weeks behind
programme. It is, in fact, slightly ahead of programme as now
adjusted following your extension of time dated... *etc. as
appropriate.*]

We shall be glad if the record could be corrected at the next
meeting. We are copying this letter to all those who attended
the meeting.

Yours faithfully,

3.02.3   Record of site meeting

Project title:	Shops & Offices Newbridge St., Borchester	Reed & Seymore Architects 12 The Broadway
Project ref: Meeting title:	456 4th site meeting	

Date: 7 October 1998
Location: Site

Those present:

S Gilbert	Client (Aqua Products Ltd)
A Morley	Main Contractor (Leavesden Barnes & Co Ltd)
G MacKay	" "
B Hunt	QS (Fussedon Knowles & Partners)
I Tegan	Struct. Engrs. (GFP & Partners)
I Hills	Architect (Reed & Seymore)
H A Hemming	Clerk of Works
F Adams	Services Engrs. (Black and Associates)

Item		Action
1.0	<u>Apologies</u> None	
2.0	<u>Minutes of last meeting</u> Agreed as correct	
3.0 3.1 3.2 3.3 3.4	<u>Contractor's Report</u> Progress is generally satisfactory Still one week behind programme due to late delivery of bricks AI 5 received and actioned Details of ironmongery revisions required in next two weeks	 Arch
4.0 4.1	<u>Clerk of Works' Report</u> Concern expressed about poor stacking of bricks	 MC
	(continue accordingly through agenda)	
9.0	<u>Any other business</u> None	
10.0	<u>Date of next meeting</u> 4 November 1989, 10.00 am on site	All

Distribution:

2	Client	1	Services Engr
3	Main Contractor	1	Clerk of Works
1	QS	1	File
1	Struct Engr		

3.03   Site diaries

You should ensure that your general foreman or agent on site, who is probably the 'person-in-charge' for the purposes of JCT 98 and WCD 98 clause 10 (IFC 98 clause 3.4, MW 98 clause 3.3), keeps a careful and detailed diary record. Some contractors have a special form for this purpose and while this does ensure that certain standard records are kept, it can be restrictive and lead to vital but non-standard information not being recorded. A sample pro-forma is shown as **document 3.03.1**.

Site diaries should record accurately any complaints or observations made by the architect or the clerk of works, consultants if appointed, quantity surveyor, building control officers, etc. Accuracy and truthfulness are important because if the diaries are later tendered in evidence, for example in an arbitration, or are used to refresh the memory of a witness, they will be rejected if they are clearly one-sided.

The following are some of the more important items which should be recorded, but the list is not exhaustive:

- Climatic conditions
- Deliveries of materials, with particulars of any defects or shortages
- Plant, with details of any breakdowns etc.
- Labour, with details of any problems, dismissals etc.
- Subcontractors' progress and problems
- Information received or outstanding
- Visitors to the site
- Progress in relation to programme

The diary should also record all other matters concerning the running of the contract on site. **Document 3.03.2** is a specimen page from a site diary.

3.03.1 Pro-forma site diary page

W.E. Build & Co Ltd **SITE AGENT'S DIARY**	

Title of contract:
Contract period: weeks from.   Week No.
Completion date: Extensions to date: weeks
DIARY FOR:

TODAY'S LABOUR   COMMENTS ON LABOUR

Foremen
Timekeeper
Gen. Lbrs
Bricklayers
Br. Lbrs
Carpenters
Joiners
Tilers/Slaters
Plasterers
Plumbers
Painters
Others

WEATHER:
  Effect on work:

ON SITE PLANT AND MATERIALS   COMMENTS

INFORMATION REQUIRED
  Details   Type   Comments

INFORMATION RECEIVED
  Details   Type   Comments

[cont.]

(3.03.1   *cont.*)

DELAYS	
Cause	Comments
Weather Other	

VISITORS	
Details	Comments

HEALTH & SAFETY	
Details	Comments

GENERAL	
Details	Comments

_____
Site Agent

To be kept in duplicate – copy to Head Office daily

3.03.2 Specimen page from site diary

Date.	Monday 17 January 2000
Weather:	Continuous rain 11 a.m. to 1 p.m. Work stopped. Otherwise dull but dry. Temperature above freezing.
Labour:	Own labour. 10 bricklayers, 5 bricklayers' labourers 12 general labourers, Block H & I. Subcontractors Dom. Roof tilers, none (failed to appear: phoned - no response. Reported H/o 10.15 a.m.) 3 chippies Block D 6 plasterers & 3 labourers Blocks B & C.
Plant:	Tower Crane: temporary b/down 2 p.m.-2.30 p.m. Etc.
Materials:	1 load 10 thou. facing bricks 9.15 a.m. Approx. 10 per cent chipped and unusable. Wrote on delivery note accordingly. Etc.
Information:	Revised Architect's drg. no. 2751/Revision/R Etc. Advised H/o 10.15 a.m. that joinery details, due today, not received.
Visitors:	8.30 a.m. Architect and QS. A instructed me orally to pull down 3 sq. m. facing brickwork Block H as pointing said unsatisfactory. Do not agree. Asked to confirm in writing. Reported H/o 10.15 am. 2.30 p.m. Bldg. Insp. Condemned stanchion base concrete Block k. Phone H/o tomorrow.
Progress:	Block H. up to programme; Block J up to programme (subject to arrival of tilers by tomorrow); Block I brickwork 1 week behind. Notified H/o by phone that more labour is needed.

3.04   The master programme

JCT 98, clause 5.3.1.2 requires you to provide the architect with two copies of your master programme as soon as possible after the signing of the contract. You must also provide him with two copies of your revised programme each time he grants an extension of time or otherwise changes the completion date.

This clause may be deleted by the employer, but even if it is you should send the architect your programme so that he may be aware of your intentions. In either case, however, the master programme is not a contract document as defined.

Often, the bills will specify the type of programme required, e.g. bar chart, network analysis, etc. Sometimes the bills go further and require you to submit your programme 'for the approval of the Architect' – this can work very much in your interest. If the architect has specifically approved your programme it becomes very difficult for him to argue when you use the programme as evidence supporting a claim for extension of time and/or money – see Chapter 5. A network programme is of great value in a claims situation as it enables you to demonstrate how critical activities have been affected. It is even more useful when used in conjunction with a bar chart.

Document 3.04.1 is suitable for use when sending the master programme and any revision of it to the architect.

There is no requirement for a master programme in the WCD, IFC or MW forms, but the contract documents may require a programme of work to be submitted and this can be made a contractual obligation.

3.04.1   Letter regarding master programme (JCT 98)

To the Architect   *Date*

Dear Sir,

[*Heading*]

In accordance with Clause 5.3.1.2 of the Contract we enclose
two copies of our master programme for the execution of the
Works. You will note that we have presented this in both bar
chart and network form. Please acknowledge receipt.
[*If the Bills required the master programme to be submitted
'for the approval of the Architect' substitute for last
sentence:* Please let us have your approval of this programme as
laid down in the Contract Bills, p.7, Item C.]

[*When sending revisions:*
In accordance with Clause 5.3.1.2 of the Contract and
following your extension of time granted on [*date*] we now
enclose two copies of our amended and revised master programme
relating to the new Completion Date.]

Yours faithfully,

3.05 Discrepancies in documents

Because of the complexity of building contract documentation, the contracts make provision for what is to happen if there are discrepancies in or divergences between documents: JCT 98, clause 2.3, WCD 98, clause 2.4, IFC 98, clause 1.4 and MW 98, clause 4.1. Basically, these are the responsibility of the architect and clause 26.2.3 of JCT 98 for example (see Chapter 5) lists this as one of the matters which may give rise to a financial claim by you.

Your basic contractual obligation is set out in clause 2.1 of JCT 98. You are to 'carry out and complete the Works in compliance with the Contract Documents'. The other forms include a similar requirement. The clauses mentioned in the previous paragraph say what is to happen *if* you find any discrepancy in, or divergence between, documents.

While this imposes no express obligation to search for and find discrepancies and divergences it is an implied obligation, since to carry out your basic contractual obligation you must have documents defining those obligations which are consistent within themselves and with each other.

Your obligation under JCT 98 clause 2.3 and the other similar clauses arises *if* you find such a discrepancy or divergence. You are to give written notice to the architect immediately (see **document 3.05.1**) and he is then to issue instructions, compliance with which may entitle you to payment, depending on their nature, and possibly to direct loss and/or expense and/or an extension of contract time.

A difficulty arises if you do not find the discrepancy at all or until it is too late for the architect to issue an instruction in time to prevent delay. In that case, you are probably entitled to appropriate extensions of time and payment under clause 26 of JCT 98 and the similar provisions in the other contracts, but possibly reduced in the light of your failure to spot the discrepancy earlier. You should resist any arguments to the contrary.

Clause 2.3 deals only with your finding a discrepancy or divergence. It says nothing expressly about what is to happen if the architect or quantity surveyor find them. Clearly, however, there is an obligation in such an event for the architect to issue appropriate instructions without being asked.

Errors in the bills of quantities prepared by the quantity surveyor are dealt with in JCT 98 clause 2.2.2. Clause 2.2.2.2 refers to 'any error in description or in quantity or omission of *items*' as well as to failure to prepare the bills in accordance with the prescribed method of measurement. Pricing errors are your own responsibility both under the general law and under clause 14.2 of JCT 98.

Although you have no express contractual obligation to notify either the architect or the quantity surveyor of any errors which you may come across, you are strongly advised to do so, particularly where the error is against you! (**Document 3.05.2.**)

Architects are not infallible and errors will sometimes occur in the documents which they provide, such as the setting-out drawings under clause 7 of JCT 98. If you discover inaccuracies you should draw the architect's attention to them at once (see **document 3.05.3**). If such drawings are unclear or need amplifying so as to enable you properly to set out, the architect is bound to give you the necessary further drawings or details under clause 5.4. WCD 98 includes a similar provision at clause 2.3.1 regarding discrepancies between the employer's requirements and the definition of the site boundary given by the employer. **Document 3.05.3** can be amended to suit.

Clause 1.4 of IFC 98 requires the architect to issue instructions in regard to the correction of the inconsistencies, errors or omissions, and if you find any errors, etc., you should draw his attention to them.

WCD 98 provides for the possibility of discrepancies within either the employer's requirements or the contractor's proposals. It is silent in respect of discrepancies between the two documents. The employer by the third recital satisfies himself that 'the Contractor's Proposals appear to meet the Employer's Requirements'. There is a presumption in footnote (b) to the recitals that the employer accepts a divergence and amends his requirements accordingly. There is no provision to cover what happens if he fails to spot a divergence but it is probable that the contractor's proposals prevail.

3.05.1 Letter where there are discrepancies or divergences in the contract documents

---

To the Architect   *Date*

Dear Sir,

[*Heading*]

On examination of the Contract Drawings and Contract Bills we find the following discrepancies:

[1. Bills, Item B, p.51, gives the roof tiling as Cosdon Watertite interlocking tiles whereas Drg No. 9257/5 (Rev) shows Beacon old hand-made clay tiles.
 2. Drg No. 9257/13 shows a recess in the north wall of Room 57 which is not shown on Drg No. 9257/19.
etc.]

Will you please issue your instructions with regard to these discrepancies as soon as possible as it will be impossible for us to order the necessary materials until these problems are resolved.

Yours faithfully,

3.05.2   Letter to architect *re* errors in contract bills

To the Architect   *Date*

Dear Sir,

[*Heading*]

On examination of the Contract Bills it is apparent to us that
Item B on p.51 relating to roof tiling is incorrect in that it
appears to give only half the total area of roof tiling
actually required.

[*Alternative:*
On examination of the Contract Bills we note that Items B to G on
p.51 are not measured in accordance with SMM 7 as required in
Clauses 2.2.2.1 of the Contract in that the following items
have not been measured: *specify.*]

[*JCT 98*: In accordance with Clause 2.2.2.2 of the Contract this
error/departure will require to be corrected and the
correction treated as a Variation under Clause 13.2.

Please confirm that this will be done and that the amount of the
correction will in due course be included in certificates when
the work is carried out.]

[*IFC 84*: Please issue your instruction under Clause 1.4.]

Yours faithfully,

3.05.3 Letter pointing out inaccuracies in setting-out dimensions

To the Architect *Date*

Dear Sir,

[*Heading*]

With reference to your Site Plan showing dimensions for
setting out the works, we have to inform you that following our
own site survey, we have found the following errors and
omissions on your drawing:
[*specify as appropriate*].

Will you please issue a revised drawing showing correct
dimensions so that we may proceed with the setting-out of the
works. Since setting-out is scheduled to commence on Monday
next it is essential that we received your revised drawing no
later than first post on the Friday beforehand.

Yours faithfully,

3.05.4 Letter notifying discrepancy within employer's requirements or contractor's proposals under WCD 98

To the Employer
or Employer's Agent]   *Date*

Dear Sir,

[*Heading*]

We have identified the following discrepancy within your
Employer's Requirements that is not dealt with by the
Contractor's Proposals:
[*description of discrepancy*]

We propose to comply with the requirement as stated in
paragraph 163 of the Employer's Requirements. If you wish us to
comply with the alternative specification in paragraph 287 of
the Employer's Requirements you should notify us accordingly
in accordance with clause 2.4.1 and this will be treated as a
change in the Employer's Requirements.

[*Alternative:*
We have identified the following discrepancy within our
Contractor's Proposals:
[*description of discrepancy*]

We propose to carry out the work in accordance with paragraph
206 of the Contractor's Proposals.

Yours faithfully,

3.06 Compliance with statutory requirements

Clause 6.1 of JCT 98 obliges you to comply with, and give all notices required by or under statutory requirements as therein defined. If compliance involves a change from what is set out in the contract documents this is to be treated as a variation.

As with discrepancies in the contract documents (see 3.05) you are under an implied obligation to find divergences between what the contract documents require and any relevant statutory requirements. If you find them you must notify the architect immediately (**document 3.06.1**). Under JCT 98 the architect *must* then issue instructions within seven days. He must also do this if he himself finds such a divergence. IFC 98 imposes no express duty on the architect to issue instructions following notification of a divergence but there is no doubt that he is bound to do so under the general law. MW 98 is similar. If these instructions result in the works being varied they are treated as a variation under clause 13.2, except where they relate to performance specified work under clause 4.2 when the contractor carries out the work as his own cost.

WCD 98 places the cost liability entirely with the contractor except where the alteration results from a change in the statutory requirements during the course of the contract.

Sometimes you will have to carry out emergency work in order to comply immediately with a statutory notice, e.g. a prohibition notice served by the Health and Safety Executive under the Health and Safety at Work etc. Act 1974. In that event you should confine yourself to carrying out only essential work and must inform the architect forthwith. If you do anything more than is strictly necessary to deal with the emergency you will not be entitled to payment for the extra work, nor will you be so entitled if you are responsible for the emergency. **Document 3.06.2** is a suitable letter.

The architect may come back and deny that there was an emergency or allege that the work you have carried out was more than necessary. You must reply according to the circumstances (**document 3.06.3**).

3.06.1 Letter regarding divergence between statutory requirements and contract documents

---

To the Architect   *Date*

Dear Sir,

[*Heading*]

We must draw your attention to a divergence between the
contract documents and statutory requirements as follows:
[*give details of the divergence*].

[*JCT 98:* May we please have your instructions regarding this
divergence as required by Clause 6.1.2 of the Contract. Unless
this instruction is received within the 7 days specified in
that clause we shall be compelled to stop work as we are then
programmed to commence construction of the walls concerned.]

[*IFC 98/MWA 98:* Please let us have your instructions by [*date*]
so that we can proceed with the work without disruption or
delay to the progress of the Works.]

Yours faithfully,

3.06.2 Letter regarding emergency work to comply with statutory requirements

---

To the Architect *Date*

Dear Sir,

[Heading]

As occupier of the site, we have received from the Council a Dangerous Structure Notice relating to the flank wall of the workshop. This notice requires us to take immediate action to ensure stability of the wall. We are now proceeding to do this by the provision of raking shores as advised by the District Surveyor.

This work constitutes emergency compliance with statutory requirements under [Clause 6.1.4 (*JCT 98*)] of the Contract and we shall therefore require the work which we are carrying out to be valued as if it were a variation instruction issued under [Clause 13.2 (*JCT 98*)].

We shall be glad to receive your immediate confirmation and any further instructions which you may consider necessary to ensure the permanent stability of the wall. [*If appropriate:* We must also give you notice that the shores which we are compelled to erect will seriously interfere with our work while they remain in position and please, therefore, treat this letter as a notice of delay under [Clause 25.2 (*JCT 98*)] and as an application in respect of the inevitable loss and expense which will arise as a result of the material effect upon regular progress under [Clause 26.1 (*JCT 98*)].

Yours faithfully,

3.06.3   Letter where the architect alleges excess work has been done

---

To the Architect   *Date*

Dear Sir,

[*Heading*]

Thank you for your letter in which you dispute that the shoring
to the flank wall of the workshop referred to in our letter of
[*date*] comprised 'such limited materials and . . . limited
work' as was reasonably necessary to secure immediate
compliance with the Dangerous Structure Notice.

We enclose a copy of a letter from the Building Control Officer
in which he confirms that the provision of the shoring was the
minimum immediately necessary to secure compliance with the
notice. You will also note that he is contemplating further
action. We have given him your telephone number and understand
that he is contacting you to discuss the situation and the
substantial further work (including under-pinning) which
will be required.

In light of this no doubt you will withdraw your objection.

Yours faithfully,

3.07   Problems with materials

Clause 8.1.1 of JCT 98 states that all materials and goods shall, so far as procurable, be of the respective kinds and standards described in the contract bills. Sometimes this is not possible for good reason. It is not sufficient for the goods and materials to be more expensive than you expected or on longer delivery. They must be genuinely not procurable by any means open to you. A good reason is that, while available at the time of tender, they are no longer available for reasons entirely beyond your control.

Document 3.07.1 is the sort of letter you must send to the architect if the specified materials are not in fact procurable. You cannot substitute goods or materials without an architect's instruction. In general, you have an absolute obligation to provide the goods and materials which you have contracted to provide and so the words 'so far as procurable' are a valuable protection for you.

JCT 98, clause 8.1.2 deals separately with your obligations as to workmanship. This must be of the standard described to the bills; if they are silent on the matter, workmanship must be of a standard appropriate to the works. If the bills provide that workmanship and materials are to be to the architect's 'reasonable satisfaction' he must then express any dissatisfaction with the quality of the workmanship and materials comprised in executed work within a 'reasonable time' – in this context 'reasonable time' is a matter of days rather than weeks.

At any time before practical completion, the architect can instruct you to open up work for inspection and to arrange for or carry out any tests of goods and materials to ensure compliance with standards: clause 8.3. You can recover the cost unless the results show non-compliance and you may also be entitled to an extension of time and reimbursement of direct loss and/or expense: see Chapter 5. IFC 98, clause 3.12 is the corresponding provision. **Document 3.07.2** is a letter you might send to the architect after inspection and testing.

The architect has extensive powers under clause 8.4 if work, materials or goods are not in accordance with the contract. He can:

- Instruct its removal or rectification; *and/or*
- After consultation with you, issue a consequential variation order (which will not entitle you to extra payment or time); *and/or*
- Issue instructions requiring opening up and testing to establish the likelihood or extent of any further non-compliance. He must have regard to the appended Code of Practice which sets out certain criteria. The contract says that, to the extent that these instructions are 'reasonable in all the circumstances to establish...' any further similar non-compliance you get no extra payment but you will be entitled to an extension of time unless the results of the inspection and testing were adverse.

WCD 98, clause 8.4 places similar powers in the hands of the employer.

You can object to the further instruction and if it is not modified or withdrawn, the dispute is referable to adjudication. **Document 3.07.3** may be appropriate.

Clause 3.13 of IFC 98 provides that *if* you discover a failure of work or of materials or goods during the carrying out of the works you must write to the architect telling him of the action you will take *immediately* at no cost to the employer to establish that there is no similar failure in work or of materials or goods which you have already executed or supplied. The obligation extends to failure due to a subcontractor's fault and, in the case of goods and materials, applies whether or not the items have already been incorporated in the works. You must notify the architect within 7 days of discovering the failure; see **document 3.07.4**.

The architect may then issue instructions requiring you to open up for inspection any other covered work or to arrange for, or carry out, tests on any other materials or goods to establish that there is no similar failure. Except where there are considerations of safety or statutory requirements (when he can issue instructions immediately) the architect must wait for up to 7 days from the discovery for you to notify him of the action you propose to take. If he is dissatisfied with your proposals or you fail to put them forward he can issue instructions to you.

You must comply with the instruction immediately even if you object to it. You must make your objection within 10 days of receipt of the instruction and you must state your reasons for objecting in writing (**document 3.07.5**) and the architect then has 7 days in which to withdraw the instruction or modify it so as to remove your objection. If he does not do so, the resulting dispute as to whether the nature or extent of the opening-up or testing was reasonable in all the circumstances is referred to an adjudicator or even directly to arbitration. If the adjudicator or arbitrator finds in your favour he will decide the amount to be paid to you by the employer and also deal with any necessary extension of time.

Clause 16 of JCT 98 and clause 1.10 of IFC 98 deal with unfixed materials and goods intended for use in the works and many legal problems arise under these provisions. They are discussed in the law books.

These provisions require that any materials delivered to the site are not to be removed unless the architect consents. It may sometimes be necessary to remove those materials for good reason: see **document 3.07.6**. See also Chapter 4 (section 4.08) as to the inclusion of the value of off-site goods and materials in interim certificates.

3.07.1   Letter to the architect where the materials are not procurable

---

To the Architect   *Date*

Dear Sir,

[*Heading*]

We regret to inform you that upon enquiring of our suppliers,
who are a national firm of builders' merchants, we have been
informed that the roof tiles specified in the Contract
Documents are no longer procurable as the manufacturer is in
liquidation. We enclose a copy of our suppplier's letter which
indicates that an acceptable alternative of similar
appearance might be . . . Unfortunately, the price of these
tiles is £[*x*] per hundred higher than the price of the
specified tiles as shown on the quotation which we obtained at
the time of tender, a copy of which is also enclosed.

May we please have your instructions under Clause [13.2 (*JCT
98*) or 3.6 (*IFC 98*)] regarding the use of an alternative tile.
Please treat this matter with urgency. As you will see from our
supplier's letter, the alternative tile which they suggest
has a delivery period of three months and if you decide to
instruct us to use this tile it will be necessary for us to place
the order within the next 7 days if delay to the progress of the
works is to be avoided.

Yours faithfully

3.07.2   Letter to architect after inspection or testing

To the Architect   *Date*
[*or Employer under WCD 98*]

Dear Sir,

[*Heading*]

Following your instruction under Clause [8.3 (*JCT 98 or WCD 98*)
or 3.3 (*IFC 98*)] requiring us to [open up [*describe work*] or
test [*describe materials*]] we confirm that [today you
inspected the work and found it satisfactory *or* the enclosed
test reports establish compliance with the contract].

Consequently, the costs involved are to be added to the
Contract Sum and as soon as possible we shall make application
to you for the appropriate extension of time in respect of the
delay [and submit our claim for reimbursement of the direct
loss and/or expense involved].

Yours faithfully,

3.07.3 Letter to architect objecting to compliance with JCT 98 clause 8.4.3 instruction

To the Architect   *Date*

Dear Sir,

[*Heading*]

We are in receipt of your instruction [*insert number*] dated [*insert date*] issued under Clause 8.4.3 of the Contract and requiring us to [*specify what instruction required*] which you say is necessary to establish that there is no further non-compliance.

We object to your instruction and consider it unreasonable because [*state reasons, referring to appended Code of Practice as necessary*, e.g. you have failed to consult with us or to endeavour to agree the amount or method of opening-up necessary, etc.]

We hope in the circumstances you will modify or withdraw this instruction and if you fail to do so we shall have no alternative but to refer matters immediately to an adjudicator.

Yours faithfully,

3.07.4 Letter informing architect of failure of work, materials or goods (IFC 98)

To the Architect   *Date*

Dear Sir,

[*Heading*]

In accordance with Clause 3.13.1 of the Contract we hereby notify you that on [*date, not more than 7 days before this letter*] we discovered a failure of [*give details*] to be in accordance with the Contract.

We propose immediately to [*state action to be taken*] to ensure that there is no similar failure in work already executed *or* materials or goods supplied by us.

Yours faithfully,

3.07.5 Letter objecting to compliance with IFC 98 clause 3.13.1 instruction

---

SPECIAL DELIVERY

To the Architect *Date*

Dear Sir,

[*Heading*]

We acknowledge receipt of your instruction [*insert number*] dated [*insert date*] issued under Clause 3.13.1 of the Contract and directing us to [*give details of work etc. required*].

We object to complying with that instruction because [*state reasons fully*] and we very much hope that you will be able to withdraw your instruction or to modify it so as to meet our objection. If you fail to do so within 7 days from receipt of this letter, the resulting dispute as to whether the nature and extent of the opening-up [*or testing*] instructed is reasonable in all the circumstances is referable to adjudication and we are fully prepared to invoke this procedure.

We are, of course, complying with your instruction in the meantime as required by the Contract.

Yours faithfully,

3.07.6   Letter requesting the architect's consent for removal of materials

---

To the Architect   *Date*

Dear Sir,

[*Heading*]

As you will be aware, a considerable quantity of copper tube was delivered to the site two weeks ago. While it is presently stored in a locked shed, and the site is nightly patrolled by guard dogs, our insurers have nevertheless advised that this and similar valuable materials should be stored at our own Head Office premises. A copy of our insurers' letter is attached from which you will see that it follows a spate of thefts from building sites in this area.

In accordance with Clause [16.1 (*JCT 98*) or 15.1 (*WCD 98*) or 1.10 (*IFC 98*)] we hereby ask you to give your consent in writing to the removal of these goods from site.

Yours faithfully,

3.08   Information necessary for the works

The sixth recital of JCT 98 allows the employer to provide you with an information release schedule which states what information the architect will release and the time of that release. This recital can be deleted if such a schedule is not to be provided. Clause 5.4.1 places an obligation on the architect to release the information in accordance with the schedule if the sixth recital is not deleted. Any failure by him in this respect is a relevant event under clause 25.4.6.1 giving you an entitlement to an extension of time. IFC 98 has similar provisions in the fourth recital and clause 1.7.1. There are no such provisions in WCD 98 or MW 98.

Clause 5.4.2 of JCT 98 and 1.7.2 of IFC 98 require the architect to provide to you, without charge, two copies of 'such further drawings or details as are reasonably necessary either to explain and amplify the Contract Drawings or to enable [you] to carry out and complete the Works in accordance with the contract'. He must do this as and when necessary in order to enable you reasonably to meet the completion date. These drawings and details include anything that is not covered by the information release schedule.

There is no obligation in either of these clauses for you to ask for this information. Under JCT 80 there was an obligation on you to make a specific application for this information in order to preserve your entitlement to an extension of time and/or reimbursement of loss and/or expense in the event of late delivery of this information. Under JCT 98 you no longer have to do this. The only obligation that is placed on you is, to the extent that it is reasonably practicable to do so, to inform the architect of the time when it is necessary for you to receive further drawings or details. This does however only apply when you are aware or have reasonable grounds for believing that the architect is not aware of the time that you need them.

In order to avoid argument as to whether the architect was aware of your requirements or not it would be as well to examine the information release schedule carefully and prepare your own list of items that you feel the schedule does not cover adequately and send this to the architect. If there is no information release schedule this is vital for the smooth running of the contract. It is not in the best interests of anyone concerned with the project for there to be arguments on extensions of time that could have been avoided with a modicum of pre-planning.

Document 3.08.1 gives a suitable covering letter when sending your own list to the architect. **Document 3.08.2** is a sample letter notifying the architect that you require information which has not been included on earlier schedules.

3.08.1   Letter notifying information requirements

To the Architect   *Date*

Dear Sir,

[*Heading*]

The Information Release Schedule does not cover the following
information that is not yet in our possession but which we
require to carry out and complete the Works:
[*Description of information*]

Would you please note that we require this information no later
than [*date*] in order to maintain our programme for the Works.

[*Alternative:*
There being no Information Release Schedule on this project we
enclose our own schedule of further information required and
the dates that the various items should be received by us in
order that we may carry out and complete the Works in
accordance with the Contract Conditions.]

Yours faithfully,

3.08.2 Letter requesting information

To the Architect   *Date*

Dear Sir,

[*Heading*]

We hereby apply for the following information [that is not
noted on the information release schedule] required by us for
the carrying out and completion of the Works:
[*Description of information*   *Date required*
Joinery details of entrance   [*date*]
screen
Colour schedule for internal paintwork   [*date*]
[*etc..*]

[*Alternative:*
We hereby apply for the supply of the Joinery details for the
entrance screen. You will see from our Master Programme, as
last revised, that we intend to commence installation of these
screens on site on [*date*] and in order to allow for the ordering
of the necessary materials and for the work of fabrication to
be carried out in our joinery workshops we shall require this
information no later than [*date.*]

Yours faithfully,

3.09   Instructions generally

The architect's powers to give you instructions are strictly limited. JCT 98, clause 4.1.1, WCD 98, clause 4.1.1 and IFC 98, clause 3.5.1, oblige you to comply only with those instructions issued by the architect (for architect read employer in the case of WCD 98) which the contract conditions specifically empower him to issue. If the architect issues an instruction which you consider or suspect is not within his powers, you must not comply with it but question it immediately using the procedure laid down in JCT 98 and WCD 98, clause 4.2, and IFC 98, clause 3.5.2. **Document 3.09.1** is a suitable form.

It is interesting to note in relation to **document 3.09.1** that where the architect, in reply, specifies a contract clause which he believes authorises his instruction you are then entitled to treat the instruction as being valid under that clause and become entitled to all the consequences which flow from that, such as extensions of contract time, and reimbursement of loss or expense. You should consider carefully whether it is in your interests to do this rather than to continue to refuse to accept the instruction and rely on your common law rights.

Should you not accept that the clause stated by the architect as his authority is such, you have the right to require immediate reference to the dispute resolution procedure under the contract.

Note that only the architect has power to issue instructions. You must not act on any 'instructions' from anyone else, e.g. a consulting engineer or even the employer himself, nor must you allow any of your subcontractors to do so.

This is one of the most common areas from which problems arise in construction contracts. If you do take instructions from someone (the employer himself for example) who is not empowered to give them and he denies that the instruction was given you do have a problem that only adjudication or arbitration will resolve.

Clause 4.3.1 of JCT 98 and WCD 98 and clause 3.5.1 of IFC 98 require all architect's instructions to be issued in writing, but clause 4.3.2 of JCT 98 provides machinery for instructions given otherwise, i.e. by word of mouth. Sensibly, IFC 98 makes no provision for oral instructions, and you should be cautious about acting on them and ask for the instruction to be put in writing. Under JCT 98 you must confirm oral instructions to the architect in writing within seven days and if he does not dissent in writing within a further seven days they take effect as instructions as from the expiry of that second seven-day period. If you do not confirm in this way but nonetheless comply with them, the architect *may* confirm them himself at a later date, but if he fails or declines to do so, you will have no remedy. **Document 3.09.2** gives alternative forms of letter for these two cases.

The clerk of works has no power to issue instructions. He is merely an *inspector* on behalf of the employer and he certainly cannot tell you how to do your work. Nevertheless, as the employer's inspector his opinion is likely to carry a little weight and you would be unwise totally to ignore what he says. If he issues any directions to you they have no effect under the contract unless they relate to matters on which the architect himself has power to issue instructions under the contract *and* the architect himself confirms them in writing within two working days of their being given. Note that it is not enough for you to confirm them to the architect as with architect's oral instructions. The architect must confirm them to you. **Document 3.09.3** is a letter you might write on receipt of directions from the clerk of works.

The Minor Works form approaches this situation in the same way as JCT 98 and IFC 98, that is, it requires any instruction to be in writing. It has no provision similar to clause 4.2 of JCT 98 but it does provide for oral instructions which must be confirmed within two days by the architect.

If the architect does not do this it is as well to remind him politely of his obligation in order to ensure that it is properly recorded and that you will be paid for the work so instructed. **Document 3.09.4** is a suggested letter.

3.09.1   Letter querying the validity of the architect's instruction

To the Architect   *Date*

Dear Sir,

[*Heading*]

We have received your AI No. 1 of [*date*] requiring us to [*give details*].

In accordance with Clause [4.2 (*JCT 98*) or 3.5.2 (*IFC 98*)] of the Contract we request you to specify in writing the provision of the Conditions which empowers the issue of this instruction.

Yours faithfully,

3.09.2   Letter regarding oral instructions

To the Architect (JCT 98/IFC 98)   *Date*
[*or Employer under WCD 98*]

Dear Sir,

[*Heading*]

On [*date*] you orally instructed us to [*specify nature of
instruction*]. [Please accept this as our confirmation of this
instruction which, if not dissented from by you in writing
within 7 days from receipt of this letter, shall thereupon take
effect. (*JCT 98/WCD 19*)]

There is no provision for instructions to be given orally in
the IFC 98 form of contract. We shall be happy to comply with
this instruction as soon as we receive confirmation in writing
from you. (IFC 98)]

[*Alternative:*
On [*date*] you orally instructed us on site to [*specify nature
of instruction*]. We duly carried out this work on [*date*] having
recorded your instruction in our site diary. Photocopies of
relevant diary extracts confirming the foregoing are
attached.

Unfortunately, however, due to an administrative error in our
office, we omitted formally to confirm the instruction to you
in writing as required by Clause 4.3.2 of the Contract. Will
you please be good enough now to confirm it in writing
yourself, as provided in Clause 4.3.2.2, so that it may take
effect as a valid instruction from the date on which it was
given to us orally.]

Yours faithfully,

3.09.3　Letter requesting confirmation of the clerk of works' directions (JCT 98)

To the Architect　　　　　　　　　　　　　　　　　　　　*Date*

Dear Sir,

[*Heading*]

We enclose copies of a number of directions which were handed to us on site today by the Clerk of Works.

As required by Clause 12 of the Contract, will you please confirm by return those directions which you wish us to treat as Architect's instructions under the Contract. As you will be aware, your confirmation is necessary within two working days of the directions having been given if they are to be treated as valid instructions.

Yours faithfully,

3.09.4 Letter requesting confirmation of architect's oral instruction (MW 98)

To the Architect/Contract Administrator   *Date*

Dear Sir,

[*Heading*]

We do not appear to have received confirmation from you in respect of the instructions you gave orally on your site visit 4 days ago.

Clause 3.5 of the Contract requires that you give written confirmation of any oral instruction within 2 days.

We look forward to receiving your written confirmation of the following instructions at your earliest convenience:
[*list*]

Yours faithfully,

3.10 Variations

In the absence of a variations clause, the architect would have no power to require you to make any changes in the work you have contracted to do. The nature of building projects is such that a variations clause is inevitable. However, JCT 98, clause 13, while appearing to be very wide in its terms, in fact limits the architect's powers considerably.

The architect can order a variation only if what he proposes falls within the definition set out in clause 13.1 and still leaves the scope and nature of the work (as defined in the contract documents) substantially unaltered. It is very difficult to say exactly where the line must be drawn. **Document 3.10.1** provides an example of what might be regarded as an extreme situation but one where the architect is clearly trying to exceed his powers.

The position is the same under IFC 98, where clause 3.6 is in terms broadly similar to JCT 98, clause 13.

The employer under WCD 98, clause 13.1, can make changes to the employer's requirements in a similar fashion to the provisions for variations under JCT 98.

3.10.1 Letter when a purported variation order changes the scope and character of the work

---

To the Architect *Date*

Dear Sir,

[*Heading*]

We have today received your AI No. 387 of [*date*] which purports to have been issued under Clause [13.2 (*JCT 98*) or 3.6 (*IFC 98*)] of the Contract. Compliance with this instruction would mean that the scope and character of the Works, as described in the Contract Documents, would be so substantially changed that they would no longer be the Works which we contracted to construct.

[*Give details, e.g.:*
Our Contract is for the construction of 12 blocks of three-storey flats of load-bearing brick construction. Your instruction seeks to convert 5 of these blocks into 7 storey buildings of reinforced concrete frame structure and further seeks to add two more blocks of similar construction. This is clearly not an alteration or modification of the design, quality or quantity of the Works as shown on the Contract Drawings and as described by or referred to in the Contract Bills. This instruction is, therefore, not one which you are empowered to issue under the Contract.]

[If this alteration is indeed essential to the Employer we shall be very pleased to enter into discussions with a view to negotiating a new contract embodying these changes. However, our programme shows that we were due to commence the construction of one of these blocks on Monday next and a decision on this matter is therefore of some urgency.]

Yours faithfully,

The example in **document 3.10.1** deals with a single major change, but the situation will also arise where so many variations are ordered that, in the end, you may find yourself being asked to do something quite different from what you have contracted to do. However, the mere number of variation orders issued has nothing to do with it – despite the fact that the relevant clauses refer to 'a Variation' rather than 'Variations'. The point is whether or not the nature of the variations has cumulatively effected a major change in the scope and nature of the works. An extreme example might be where you have contracted to build a factory/workshop and by a series of VOs you find that it has gradually been changed into an office building.

If you suspect this sort of thing is happening, you should seek specialist advice. In general, you are contractually obliged to comply with *all* instructions which the architect is empowered to issue, including variations. However, there is one kind of variation with which you are not obliged to comply if you have a 'reasonable objection' to doing so (a right given by clause 4.1.1.1 of JCT 98 and clause 13.5.1 of IFC 98), namely a variation in 'any obligations or restrictions imposed by the Employer in the Contract Bills in regard to' those matters which are listed in clause 13.1.2 of JCT 98 and clause 3.6.2 of IFC 98.

Those matters are:

- Access to the site or use of any specific parts of it
- Limitations of working space or working hours
- The execution or completion of the works in any specific order

A similar right to object is given by clause 4.1.1 of WCD 98 in respect of employers' instructions under clause 12.1.2.

Although SMM 7 permits a substantial number of restrictions and obligations to be imposed on you through the medium of the contract bills (provided they do not seek to over-ride or modify the application or interpretation of the contract as printed) the architect's power to vary them is limited to those which are listed in clause 13.1.2 of JCT 98 or clause 3.6.2 of IFC 98 and no others.

The question of what is a reasonable objection is a matter of objective judgment and in a case of dispute the problem can be referred immediately under the dispute resolution procedures in the contract. It is not sufficient simply to say that it is difficult for you to comply – you are expected to overcome those difficulties and the questions of cost and time are dealt with in the contract. Compliance with the instruction must make the continued execution of the work virtually impossible (**document 3.10.2**).

3.10.2 Objection to a variation under clause 13.1.2 (JCT 98)

To the Architect   *Date*

Dear Sir,

[*Heading*]

We are in receipt of your AI No. 154 of [*date*], Item 57 of which
seeks further to restrict our means of access to the site
beyond the restrictions already set out in the Contract Bills,
page 10, Item E.

As provided in Clause [4.4.1 (*JCT 98*) or 3.5.1 (*IFC 98*)] of the
Contract we have a reasonable objection to compliance with
this instruction in that this further restriction of access
will make it impossible for the structural steelwork to be
delivered to the site.

Will you please reconsider this instruction and either
withdraw it or issue fresh instructions which will overcome
this problem. We must request a reply within 7 days of today's
date, failing which we will regard the instruction as invalid.

Yours faithfully,

There are now two alternative procedures for the valuation of variations in JCT 98, IFC 98 and WCD 98. Alternative B is the procedure that was the only one available prior to the 1998 editions of these forms, which is a valuation made by the Quantity Surveyor (WCD is actually silent as to who does the valuation).

Under Alternative B the valuation of variations under JCT 98 and IFC 98 is a matter for the quantity surveyor and the rules of measurement and valuation are set out in JCT 98 clause 13.5 and clause 3.7 of IFC 98. Valuation of varied work will generally be on a basis of measurement and is related to the prices in the contract bills. The valuation will, therefore, take account of such items as overheads and profit which would usually be included in the bill or other prices.

Application of the bill prices or those in the IFC 98 priced document depends on whether the character of the work, the conditions under which it is to be carried out and the quantity of work involved are unchanged. While the contract does not specially require you to notify the quantity surveyor where, in your opinion, any of these factors is changed, it is sensible to do so as the quantity surveyor cannot be expected to be aware of all matters which could affect the pricing. **Document 3.10.3** is an example of the sort of letter you might send.

Alternative A is a new introduction of JCT 98, IFC 98 and WCD 98. It introduces a provision for you to prepare a 'Price Statement' in respect of any instruction that may be given.

The price statement states your price for the work based on the provisions of the valuation rules in the contract and may be accompanied by your requirements in respect of direct loss and/or expense and adjustment of time for the completion of the works. **Document 3.10.4** gives a sample letter. The price statement can be accepted and it then becomes an adjustment to the contract sum. If it is not accepted the quantity surveyor (under JCT 98 and IFC 98) or the employer (under WCD 98) is required to give reasons for the non-acceptance and supply an amended price statement. If you do not accept this amended price statement you, or the employer, can immediately refer this as a dispute to an adjudicator. Where the quantity surveyor makes no response to your price statement, clause 13.4.1.2 A5 (JCT 98) allows you to refer this to an adjudicator as a dispute. It would be unfortunate for this to be done if the failure to make a response is innocent. **Document 3.10.5** sets out alternative letters to cover these two situations.

3.10.3 Letter where a variation involves altered character, conditions or quantity of work

---

To the Quantity Surveyor   *Date*
(Copy to the Architect)

Dear Sir,

[Heading]

With reference to AI No. 71 we must point out that while the work involved is of similar character to that set out on pages 35 and 36 of the Contract Bills, the conditions under which we will have to carry out this work are substantially different from those under which the work in the Bills was to be carried out in that *[specify changed conditions, e.g.* it will be necessary for us to manhandle the materials involved into the partially-completed building since the tower crane has now been removed in accordance with our programmed intentions]*.

Will you please, therefore, make fair allowance for all this difference in the rates for this work when making your valuation.

Yours faithfully,

3.10.4   Letter enclosing a price statement

To the Quantity Surveyor   *Date*
[or Employer under WCD 98]

Dear Sir,

[Heading]

We are in receipt of AI No. 20 which requires the postponement
of the commencement of Block B. We are pleased to enclose our
Price Statement under [Alternative A of clause 13.4.1.2 (*JCT
98*), Option A of clause 3.7.1.2 (*IFC 98*)].

In accordance with that clause we also include our
requirements in respect of an amount to be paid in lieu of any
ascertainments under clause 2.6.1 of direct loss and/or
expense and also in respect of an adjustment to the time for the
completion of the works.

Yours faithfully,

3.10.5 Response when price statement not accepted or not responded to

---

To the Architect   *Date*
[*or Employer under WCD 98*]
(*Copy to Quantity Surveyor*)

Dear Sir,

[*Heading*]

We are in receipt of [the Quantity Surveyor's] notification
that our Price Statement of [*date*] in respect of your AI No. 20
is not accepted.

We do not accept [the Quantity Surveyor's] amended Price
Statement attached to that notification. We hereby give
notice of our intention to submit this dispute to
adjudication.

Unless you agree to the appointment of Mr F. Jones as
adjudicator by return of fax we shall apply to the President of
the Royal Institution of Chartered Surveyors who is the
nominator of Adjudicator in the Contract.

[*Alternative:*
We must inform you that [the Quantity Surveyor] has failed to
notify us regarding our Price Statement of [*date*] in respect of
your AI No. 20.

It is unusual for [the Quantity Surveyor] to fail in his duties
in this way and we are sure that it has happened innocently. We
are however concerned that your position regarding this price
statement should be made known as soon as possible.

Clause [13.4.1.2 A5 (*JCT 98*) or 3.7.1.2 A5 (*IFC 98*) or 12.4.2 A5
(*WCD 98*) allows us to take this matter to immediate
adjudication but we would not want to do this without your
having the opportunity to address this situation.]

Yours faithfully,

Clause 13.5 requires the quantity surveyor to take a number of matters into account and one of these which is often missed is 'liabilities directly associated with a Variation' (clause 13.5.7 of JCT 98 and clause 3.7.5 of IFC 98). An example of a situation which might arise and which ought to be dealt with under this provision is given in **document 3.10.6** and again it is sensible to draw the quantity surveyor's attention to such matters.

The basic situation is that the quantity surveyor should deal with all matters relating to the variation except its 'material effect upon the regular progress of the Works' which is primarily a matter for the architect under clause 26 and in respect of which you must make a written application to him (see **document 5.09.1**).

Under MW 98 there are no valuation rules; all that is provided for, in clause 3.6, is the valuation by the architect/contract administrator on a 'fair and reasonable basis' of any 'addition to or omission from or other change in the Works or the order or period in which they are to be carried out'. WCD 98 does have valuation rules in clause 12.5 but they are less detailed than those in JCT 98, there being no contract bills to which the price of the variation can be related, the pricing document being the contract sum analysis.

3.10.6   Letter regarding redundant materials

To the Quantity Surveyor   *Date*
(Copy to the Architect)

Dear Sir,

[*Heading*]

We have now received AI No. 227 from the Architect in which he
instructs us to substitute Welsh slates for the roof tiles
specified in the Contract Documents. Unfortunately, we had
already ordered the roof tiles from our supplier since the
period of delivery was such that this order had to be placed
four weeks ago if sufficient tiles were to be available for us
to start the tiling work on the date shown in our programme.
Copies of the quotation and order are attached.

We have telephoned our suppliers who say that the cancellation
of the order will entail a 25 per cent restocking charge. Since
this is clearly a liability directly associated with this
variation, no doubt you will make the necessary allowance in
your valuation as required by Clause [13.5.7 (*JCT 98*) or 3.7.5
(*IFC 98*)] of the Contract.

Yours faithfully,

3.11   Work done by others

Clause 29 of JCT 98 and WCD 98 and clause 3.11 of IFC 98 cover the situation where the employer wishes to carry out work, or have it carried out, while your contract work is still proceeding. It is wide in its scope. If you have been given adequate notice in the contract documents of the kind of work which is to be done you must allow it to be carried out and will be deemed to have made all appropriate allowances in the contract sum. If, however, the employer wishes to introduce work of which you have not been so notified, or which is not adequately described in the contract documents, he must seek your permission, which must not be unreasonably withheld. Any necessary extensions of time and additional cost to you can be dealt with under clauses 25 and 26 of JCT 98 and WCD 98 and clause 4.11 of IFC 98. You are entitled to be given very full information in the contract documents, sufficient for you to make all the necessary allowances as to time and cost, and the work concerned must not form part of the work which you have contracted to do. **Document 3.11.1** is an example of a letter which you might write if insufficient information has been given in the contract documents.

Both clauses relate only to work 'which is to be carried out by the Employer himself or by persons employed or [otherwise] engaged by him'. It therefore does not cover work carried out by a local authority or statutory undertaker which falls within its statutory obligations. This exclusion is in fact very limited since statutory undertakers are normally only obliged by their governing statutes to carry out work of a very limited nature, e.g. electricity boards are normally only obliged to install mains supplies to new buildings where there is an existing main of adequate capacity within a fairly short distance. While only they can install new mains or divert existing mains etc., they can refuse to do so and are only obliged to carry out the work where it falls within the statutory limits. The same considerations apply to extensions of time in respect of their work under clause 25.4.11 (JCT 98 and WCD 98) or clause 2.4.13 (IFC 98).

In general, statutory authorities carrying out works beyond their limited statutory obligation, will be deemed to be persons employed or otherwise engaged by the employer, unless of course they do it under subcontract to you.

3.11.1 Letter where the employer seeks to do work which is insufficiently identified in contract bills

To the Employer *Date*
(Copy to the Architect)

Dear sir,

[*Heading*]

We are in receipt of your letter of [*date*] in which you state
your intention to employ Sir Herbert Gussett, RA, to paint
murals in the Board Room on the 4th floor of H Block and further
that he hopes to commence work on [*date*] which is one month
before completion of this block is due in accordance with our
programme.

[Item H on p. 8 of the Contract Bills (*JCT 98 or IFC 98*) or item 63
of the employer's requirements (*WCD 98*)] refers only to your
intention to employ others to execute special finishings in
this Block. We do not consider that this bald statement
provides sufficient information to us to have made allowance
for the work which you now propose, particularly as we
understand that Sir Herbert requires special and very
complicated staging to be erected by a firm called
Michelangelo (Artists Services) Ltd.

In the circumstances we consider that this work falls within
Clause [29.2 (*JCT98 or WCD 98*) or 3.11 (*IFC 98*)] of the Contract
for which our consent is required. We are prepared to give that
consent only on the understanding that the resulting delay
will be the subject of the appropriate extension of time under
Clause [25 (*JCT 98 or WCD 98*) or 2.3 (*IFC 98*)] and the loss and/
or expense directly resulting from the inevitable material
effect on regular progress of our work will be reimbursed to us
under Clause [26 (*JCT 98 or WCD 98*) or 4.11 (*IFC 98*)].

We are sending the appropriate formal notices to the Architect
but copies are enclosed for your information.

Yours faithfully,

3.12   Other matters

There are many other problem areas and times when you must send letters or serve notices. To include them all would make this book unwieldy. You are referred to the table in Chapter 1 for a comprehensive list.

Chapter 4
Certificates and Payments

4.01 Introduction

The basic scheme for payment remains the same in JCT 98 as it did in JCT 80. There is no requirement that you make applications for interim payment. Clause 30 lays down the scheme of payment and in essence that scheme is straightforward. It depends on the architect issuing interim certificates at specified intervals – usually one month – and the employer paying you the amount shown in the certificate within 14 days from the date of issue.

The machinery for payments is the same under IFC 98. Unless otherwise stated in the appendix, the architect must issue monthly interim certificates and the employer must pay the amount certified within 14 days of the date of the certificate (clause 4.2). In both cases, this is subject to any right he may have to deduct, e.g. liquidated damages.

Under MW 98 the architect must again certify progress payments due to you at four weekly intervals. The employer must, of course, pay you within 14 days of the date of the certificate.

As far as WCD is concerned there are two alternative schemes of payment. The first, alternative A, relates to stage payments which are based on the cumulative value of the works at the relevant stage. The second, alternative B, periodic payments, is a similar requirement to that in JCT 98.

4.02 Application for payment

A new provision introduced into JCT 98, clause 30.1.2.2, and IFC 98, clause 4.2(c), is a formal right for you to make an application for payment not later than seven days before the date of an interim certificate. This application is submitted to the quantity surveyor who is obliged, if this is done, to make an interim valuation. If the quantity surveyor disagrees with the gross valuation in your application he is obliged to submit a statement to you, in similar detail to that given in your application, identifying where he disagrees with you.

This is not especially different from the procedure that was customary before the 1998 suite of JCT contracts was issued but it does put a considerable onus on the quantity surveyor to be open in respect of any adjustments that he makes to your valuation.

WCD 98 includes a similar provision where alternative B, periodic payments, applies but, rather than a right which does not have to be exercised, as in JCT 98 and IFC 98, in WCD 98 it is an obligation for you to make an application for interim payment at the period stated in the appendix.

MW 98 says nothing other than requiring the architect/contract administrator to issue certificates at four weekly intervals. The usual practice is to make an application for consideration before each certificate is due. **Documents 4.02.1–3** give some suggested wording for the covering letter to the applications for payment under the various forms. These letters include an aide memoire to the architect to remind the employer to give the written notice that is required in addition to the certificate.

The JCT contracts comply with the requirements of the Housing Grants, Construction and Regeneration Act 1996.

In JCT 98, IFC 98 and MW 98 the date of the architect's certificate becomes the 'due date for payment'. The 'final date for payment' is a date 14 days after the date of the certificate. The payment does not have to be made on the due date for payment but it must be made no later than the final date for payment. In addition to the architect's certificate, the employer has to issue a written notice of the amount due specifying the amount of the payment to be made, to what the amount of the payment relates and the basis on which that amount is calculated. This must be done within five days of the issue of the certificate.

In WCD 98 the absence of an architect creates a different regime. The due date for payment is effectively the date of the contractor's application made under clause 30.3.1.1 or 30.3.1.2. The employer has to issue his written notice of the amount due in the same format as is required by JCT 98 etc. no later than five days after receipt of the application. The final date for payment is a date 14 days after the receipt by the employer of the contractor's application.

This written notice is a very important document particularly for the employer. If the employer does not issue this notice he must pay either the amount stated in the interim certificate (JCT 98, clause 30.1.1.5; IFC 98, clause 4.3(d) and MW 98, clause 4.5.1.4) or the amount that has been applied for in the contractor's application for payment (WCD 98, clause 30.3.5).

4.02.1   Letter enclosing interim application for payment (JCT 98, IFC 98)

To the Quantity Surveyor   *Date*
(Copy to Architect)

Dear Sir,

[Heading]

In anticipation of the issue by the Architect of Certificate No. 3 we enclose for your attention our application setting out what we consider to be the amount of the gross valuation pursuant to [Clause 30.2 (*JCT 98*) or Clauses 4.2.1 and 4.2.2 (*IFC 98*)].

We look forward to receiving your valuation, the Architect's certificate and the Employer's written notice under Clause [30.1.1.3 (*JCT 98*) or 4.3 (b) (*IFC 98*)] in the same amount as we have claimed. Should there be any aspects of our gross valuation with which you disagree we have no doubt that you will let us have your statement identifying the points disagreed with, in similar detail to our application, in accordance with Clause [30.1.2.2 (*JCT 98*) or 4.2 (c) (*IFC 98*)].

Yours faithfully,

4.02.2   Letter enclosing application for interim payment (WCD 98)

To the Employer   *Date*

Dear Sir,

[*Heading*]

In accordance with Clause 30.3.1.2 of the Contract we enclose
our Application for Interim Payment No. 3. We look forward to
receiving your written notice as required by Clause 30.3.3
specifying the amount of payment proposed to be made, the basis
on which this amount is calculated and to what the amount
relates, within the next five days. We also look forward to
receiving your payment no later than the final date of payment.

Yours faithfully,

4.02.3   Letter enclosing interim application for payment (MW 98)

To the Architect   *Date*

Dear Sir,

[*Heading*]

Under Clause 4.2.1 of the Contract we hereby make application for a progress payment of the amount due to us for the four weeks ending today and which we have calculated as being £x on the basis of the value of the works properly executed [*if appropriate* and the amount agreed for variations under Clause 3.6, *etc.*].

In support of our application we enclose the following documents:
[*List*]

We look forward to receiving your certificate by return and the employer's written notice under clause 4.4.1 not later than 5 days thereafter.

Yours faithfully,

4.03   Late payment

Clause 30.1.1.1 of JCT 98 and similar clauses in the other JCT contracts reflect the Late Payment of Commercial Debts (Interest) Act 1998. This clause provides that if you are not paid by the 'final date for payment' you are entitled to the payment of simple interest at the rate of 5% over the Base Rate of the Bank of England. Document 4.03.1 is a suitable letter notifying the employer of this situation. This right to interest is additional to your right to suspend performance given by clause 30.1.4 of JCT 98, and similar clauses in the other contracts, which is dealt with in section 4.06.

4.03.1   Letter informing employer that interest is payable

To the Employer   *Date*
(*Copy to the Architect*)

Dear Sir,

[*Heading*]

We have to inform you that as we have not received payment in
respect of Certificate No. 6. Under the terms of Clause
[30.1.1.1 (*JCT 98*) or 4.2(a) (*IFC 98*) or 30.3.7 (*WCD 98*) or
4.2.2 (*MW 98*)] simple interest at 5% over the current Base Rate
of the Bank of England [a total of *x*%] applies to the
outstanding amount.

Would you please ensure that when the cheque is issued the
appropriate amount of interest is added to it.

We are not exercising our right of suspension under Clause
[30.1.4 (*JCT 98*) or 4.4A (*IFC 98*) or 30.3.8 (*WCD 98*) or 4.8 (*MW
98*)] at this time but this letter should not be taken in any way
as waiving that right.

Yours faithfully,

4.04   The withholding notice

The employer is contractually bound to pay the amount that is set out in his written notice or if he has not issued such a notice, the amount in the architect's certificate, or the contractor's application for payment under WCD 98. The only way in which he can avoid paying the full amount without penalty is to issue a 'withholding notice' (clause 30.1.1.4, JCT 98, clause 4.2.3(b), IFC 98, clause 30.3.4 WCD 98 and clause 4.4.2 MW 98). The withholding notice must be issued no later than five days before the final date for payment. There is nothing in the conditions to prevent the employer from setting out the withholding notice within his written notice setting out the amount due. As discussed above, if he does not issue the written notice he is bound to pay the amount due in full and thus loses his right to issue a withholding notice. **Document 4.04.1** sets out a letter which could be written if the employer does not issue a written notice setting out the amount due and then issues a withholding notice.

4.04.1   Where withholding notice issued incorrectly

To the Employer   *[Date]*

[Heading]

Dear Sir,

We are in receipt of your purported withholding notice in relation to [Certificate No. [*number*] (*JCT 98*, *IFC 98 and MW 98*)] [our Application for Payment No. 4 (*WCD 98*)].

We would remind you that you did not issue the written notice that is required by Clause [30.1.1.3 (*JCT 98*) or 4.2.3(a) (*IFC 98*) or 30.3.3 (*WCD 98*) or 4.4.1 (*MW 98*)] of the Conditions. In the absence of a written notice from you, Clause [30.1.1.5 (*JCT 98*) or 4.2.3(c) (*IFC 98*) or 30.3.5 (*WCD 98*) or 4.4.3 (*MW 98*)] requires that you pay the amount [*certified/applied for*] in full.

We look forward to receiving the full amount [*certified/ applied for*] not later than the final date for payment.

Yours faithfully,

4.05   Wrongful deductions from the certificate by the employer

The contracts are specific as to what sums may be deducted from amounts shown as due in interim certificates. Such deductions can be made only by the employer and not by the architect.

Under JCT 98 the employer may make deductions in respect of:

- Clause 4.1.2 – Costs incurred in employing other persons to carry out architect's instructions
- Clause 21.1.3 – Insurance premiums which you have or any subcontractor has failed to pay
- Clause 22A.2 – Unpaid work insurance premiums
- Clause 22FC.3.2 – Costs incurred by employer in employing other persons to carry out Joint Fire Code remedial measures
- Clause 24.2.1 – Liquidated damages
- Clause 27.5.4 – Following determination of your employment under clause 27.3.4
- Clause 31.3.2 – Statutory tax deductions
- Clause 35.13.5.3 – Direct payments to nominated subcontractors

Other than the last item relating to nominated subcontractors, the clause references in WCD 98 are the same (clause 4.1.2 relates to employer's instructions of course).

IFC 98 permits the employer to make the following deductions:

- Clause 2.7 – Liquidated damages
- Clause 3.5.1 – Costs incurred in employing others to carry out architect's instructions
- Clauses 6.2.3 and 6.3A.2 – Unpaid insurance premiums
- Clause 6.3 FC.3.2 – Costs incurred by employer in employing other persons to carry out Joint Fire Code remedial measures
- Clause 7.5.4 – Following determination of your employment under clause 7.3.4

In MW 98 the following deductions are allowed:

- Clause 2.3 – Liquidated damages
- Clause 3.5 – Costs of employing other persons to carry out architect's instructions
- Clause 7.2.3 – Following determination of your employment under clauses 7.2.1 or 7.2.2

These are all matters that should be set out in a withholding notice. Employers sometimes misunderstand the position and deduct from interim payments alleged set-offs and counterclaims. They may not be entitled to do this – although they may be so entitled in the case of the final payment to be made after the issue of the final certificate.

If deductions are made in a withholding notice that you consider to be wrongful, you should write to the employer in terms such as those in **document 4.05.1**.

4.05.1   Letter where deductions are wrongfully made by the employer

---

SPECIAL DELIVERY

To the Employer   *Date*
(Copy to the Architect)

Dear Sir,

[Heading]

We acknowledge receipt of your withholding notice in respect
of Interim Certificate No. [*insert number*] of [*date*], which
shows an amount due of £[*y*]. This notice states that you will
deduct the sum of £[*z*] from the certified amount in respect of
[*specify*].

We must point out to you that the terms of the Contract between
us do not permit such alleged set-offs to be made from Interim
Payments certified by the Architect and we must therefore
request that you withdraw your withholding notice in respect
of these items.

If you do not do this we shall issue a notice of intention to
refer this dispute to adjudication and avail ourselves of our
other legal rights and remedies.

Yours faithfully,

4.06 Right of suspension of obligations by contractor

The Housing Grants, Construction and Regeneration Act (1996) introduced the right to suspend work if an amount that is due under the contract is not paid in full. The amount due under the contract is either

- the amount set out in the employer's written notice as required by the contract, *or*
- the amount certified (applied for under WCD 98) when there is no written notice as required by the contract, *or*
- the amount in the written notice reduced by any amount in respect of which there is a properly issued withholding notice.

When an amount due is not paid in full you have the right, if you wish to exercise it, to suspend performance. You cannot do this without first giving proper written notice. **Document 4.06.1** shows a suitable form for such a notice.

The contracts give no specific requirements regarding the way in which this notice should be given to the employer. This is rather different from the way in which a notice of determination under clause 28 has to be given (see Chapter 7). It would be as well to use similar methods and ensure that the notice is given by actual delivery (by hand) or by special delivery. Clause 1.7 of JCT 98 notes that service can be by any effective means and the use of facsimile transmission followed by special delivery is probably best.

Suspension of performance is not a real solution and should be used only as a last resort. A notice of adjudication is often to be preferred as a means of ensuring that you receive the money due to you.

4.06.1   Letter notifying intention to suspend performance of obligations

---

SPECIAL DELIVERY

To the Employer   [*Date*]
(Copy to Architect)

Dear Sir,

[*Heading*]

We hereby notify you in accordance with Clause [30.1.4 (*JCT 98*
or 4.4A (*IFC 98*) or 30.3.8 (*WCD 98*) or 4.8 (*MW 98*)] that, due to
your failure to make payment in full of the amounts due to us
under this contract, unless we receive that payment in full we
shall suspend performance of our obligations under the
contract on [*a date 7 days after the giving of the notice*].

If we do suspend performance we shall recommence performance
of our obligations a reasonable time after payment is received
in full.

We would remind you that any period of suspension for this
cause gives us an entitlement to an extension of time for
completion of the Works under Clause [25.4.18 (*JCT 98*) or
2.4.18 (*IFC 98*) or 25.4.17 (*WCD 98*) or 2.2 (*MW 98*)] and also an
entitlement to the reimbursement of loss and/or expense
caused thereby.

We have no desire to see the project delayed for this reason and
would hope to receive payment in full before this notice
expires.

Yours faithfully,

4.07 Problems with interim certificates

The contracts do not make any specific provision for what is to happen if the architect fails to issue an interim certificate at the proper time. You cannot determine your employment under clause 28 or clause 7.5 (see Chapter 7) on this ground since the right to determine only arises if the employer fails to honour a certificate which has actually been issued or if he himself interferes with or obstructs its issue.

The payment provisions of JCT 98, IFC 98 and MW 98 rely on the architect's certificate being issued. A failure to issue a certificate in accordance with the conditions is a breach of contract and is a matter that can be referred immediately to adjudication.

If the architect fails to issue an interim certificate, you must take legal advice, but in the meantime you should write to the architect (**document 4.07.1**) by fax and special delivery and send a copy of your letter to the employer, since there is case law which suggests that the architect's failure to issue certificates is a default for which the employer is responsible in law.

Under WCD 98 there is no certification process and the absence of the employer's written notice is provided for in the contract and has been discussed earlier.

4.07.1   Letter where no interim certificate is issued

SPECIAL DELIVERY

To the Architect   *Date*
(Copy to the Employer)

Dear Sir,

[Heading]

We do not appear to have received a copy of Interim Certificate
No. [x] which was due to be issued on [*specify date*] in
accordance with the terms of Clause 30.1.1.1 (*JCT 98]* or 4.2
(*IFC 98* or 4.2.1 (*MW 98*)] of the Contract. We shall be grateful
if you will confirm that the certificate was so issued and
forward a copy to us immediately.

Failing receipt of such confirmation within 7 days from the
date of this letter we shall treat this as a dispute or
difference that can be referred to an adjudicator and shall be
compelled to give notice of intention to refer the dispute to
adjudication.

We are copying this letter to the Employer for his information.

Yours faithfully,

4.08 Off-site goods and materials

Under JCT 98 (clause 30.3), IFC 98 (clause 4.2.1(c) and WCD 98 (clauses 15.2 and 30.2B.1.3), materials off site will not be paid for unless they are 'listed items'. That means that the list must have been provided by the employer and annexed to the contract bills, the specification, the employer's requirements etc. and thereafter included in the contract documents.

There is no provision for materials other than those listed by the employer in the contract documents to be paid for before delivery to site.

There are certain conditions in addition to being listed that must be met before payment is made for materials before delivery to site:

- The listed items must be in accordance with the contract.
- They have been set apart at the premises where they have been manufactured or stored, clearly and visibly marked, individually or in sets, to identify them as being held to the order of the employer. They must be marked so as to make clear that they are destined for the works.
- You must provide reasonable proof that the property in the goods has been passed to you and that the other conditions are satisfied as well as adequate proof of insurance cover.

If so required in the appendix to the contract you will have to provide a bond in respect of the sums paid for the listed materials. This bond will either be in the form set out in the contract or the employer may draft a bond specifically for the contract in question (see Chapter 2.13).

MW 98 has no provision for payment in this respect.

Document 4.08.1 is a letter asking for listed items to be paid for.

4.08.1 Letter requesting the inclusion of the value of listed items in an interim certificate

---

To the Architect (JCT 98, IFC 98)   Date
(or Employer under WCD 98)

Dear Sir,

[Heading]

We shall be grateful if you will include the value of the following items that were listed [in the Contract Bills/Specification/Employer's Requirements] to be paid for before delivery to site in the next Interim Certificate under the Contract under the provisions of Clause [30.3 (*JCT 98*) or 4.2.1(c) (*IFC 98*) or 15.2 and 30.2B.1.3 (*WCD 98*)]. The goods and materials are: [*specify*].

We confirm that the relevant provisions of the above mentioned clause(s) have been complied with, and the goods are available for your inspection at [*specify*]. We further confirm that the goods and materials are our property and enclose proof of ownership (*e.g.* copy of sale contract], and confirm that they are insured in accordance with the provisions of the contract.

Yours faithfully,

4.09   Retention

Clause 30.5.1 of JCT 98 (30.4.2.1 of WCD 98 and 4.4 of IFC 98) makes the employer a trustee of the retention fund both for you and for nominated subcontractors. This is to protect your interests should the employer become insolvent.

Under JCT 98 or WCD 98 you or any nominated subcontractor can ask a private employer – not a local authority employer – to put the retention money in a separate bank account designated to show that the retention is trust money and therefore not available to other creditors should the employer become insolvent.

All that is required is a letter from you requesting the employer to place in a separate bank account the amounts of retention which are going to be deducted under interim certificates. The wording of the relevant clause (30.5.3 of JCT 98 and 30.4.2.2 of WCD 98) seems quite clear and any requirement that you make a separate request in respect of each payment is onerous and should be resisted. No matter when the request is made, or by whom, the whole of the retention fund held by the employer should be placed in the separate account.

Under IFC 98 the status of the retention money is dealt with by clause 4.4. The employer's interest in the retention money is only fiduciary as a trustee for you if the employer is not a local authority. In the case of a private employer there is always some risk of future insolvency and you can ask for the retention to be set aside in a special bank account even though there is no such specific provision in the contract.

However, even though the retention is trust money and may be placed in a trust account, it is nevertheless primarily intended as the employer's security for proper performance of the contract by you and your subcontractors alike. Clauses 30.1.1.2 (JCT 98), 30.4.3 (WCD 98) and 4.4 (IFC 98) empower the employer to have recourse to the retention to satisfy his own rights of deduction under the contract.

If you decide that a separate trust account is desirable, a suitable form of request is found in **documents 4.09.1** and **4.09.2**.

There are no such provisions in MW 98, clause 4.2 of which deals with retention under that form.

4.09.1   Letter requesting the employer to open a trust account for retention

To the Employer   *Date*
(Copy to the Architect (JCT 98))

Dear Sir,

[Heading]

Pursuant to Clause [30.5.3 *(JCT 98)*, 30.4.2.2 *(WCD 98)*] of the
Conditions of Contract we hereby request you to place the
Retention in respect of the next and all subsequent [Interim
Certificates/payments] in a separate banking account so
designated as to identify the amount held in the account as the
Retention held by you on trust as provided in Clause [30.5.1
(JCT 98), 30.4.2.1 *(WCD 98)*] of the Contract Conditions.

Please certify [to the Architect, with a copy] to us, that the
Retention in respect of each certificate has been so placed.

As soon as you have opened this account, we shall be grateful if
you will inform us of the Bank, the account number and its
designation.

Yours faithfully,

4.09.2 Letter requesting the employer to open a trust account for retention (IFC 98)

To the Employer   *Date*
(Copy to the Architect)

Dear Sir,

[*Heading*]

We hereby request you to place the Retention money in the next and all subsequent Interim Certificates under the Contract in a separate bank account set up for that purpose and duly identified as trust money as specified in Clause 4.4 of the Contract.

As soon as you have opened this account, we shall be grateful if you will inform us of the Bank, the account number and its designation,

Yours faithfully,

4.10   The Construction Industry Scheme

Amendment 1 issued by the JCT in June 1999 applies to all versions of the JCT contract. This amendment deals with the introduction of the Construction Industry Scheme (CIS) which is the regime which applied from 1 August 1999 onwards to any payment made by a 'Contractor' to a 'Subcontractor'. The crucial aspect of the regulations introduced by the CIS is that any payment to a subcontractor may only be made if the subcontractor is in possession of a valid registration card or tax certificate. There are substantial sanctions where payments are made in the absence of the necessary registration card or tax certificate including a fine of up to £3000 and reimbursement to the Inland Revenue of the tax that should have been deducted. The contractor making such payments also runs the risk of having his own tax certificate withdrawn.

For the purposes of the CIS an employer under a building contract will be a 'Contractor' if the overall value of the construction operations being carried out exceeds an average of £1m per annum.

The detail of the operation of these provisions is a matter for accounts and taxation specialists but the important thing as far as the JCT contracts are concerned is to ensure that the appropriate authorisation required by clause 31.3 has been provided to any employer who is defined as a 'Contractor' for the purposes of the CIS.

Chapter 5
Delays and Disruptions

5.01 General

Four clauses in JCT 98 cause special difficulties in practice. They are:

- Clause 23 – Date of possession, completion and postponement
- Clause 24 – Damages for non-completion
- Clause 25 – Extension of time
- Clause 26 – Loss and expense caused by matters materially affecting regular progress of the Works.

Briefly explained, the substance of these clauses is as follows. The employer is obliged to give you possession of the site on a stated date and you are then required to commence the Works and complete them *on or before* a further stated date (clause 23.1.1). If clause 23.1.2 is stated in the appendix to apply, the employer may delay giving possession of the site to you for a period of not more than six weeks, but in these circumstances you will be entitled to an extension of time and be able to claim any 'direct loss and/or expense' which you incur as a result of possession being deferred. Apart from this provision, failure to give you possession on the due date is a breach of contract and the architect has no power to extend the contract period. The architect is given the power to postpone any *work* to be executed under the contract. The contract draws a distinction between 'work' and 'the Works', but it seems clear that the architect's power extends to postponement of the whole of 'the Works' as well as to any part of them (clause 23.2).

If matters arise which delay you in carrying out the Works you must give notice to the architect immediately. Certain matters will then entitle you to have the Completion Date of the Works postponed (clause 25). Some of these matters also entitle you to recover any 'direct loss and/or expense' which you may incur as a result of their material effect on regular progress of the Works (clause 26).

If you fail to complete the Works by the original date or any later date fixed by the architect under clause 25, the architect must issue a certificate to that effect and the employer then becomes entitled (after due notice to you) to recover 'liquidated and ascertained damages' which he may deduct from any payments due to you or may recover from you as a debt (clause 24).

The position is similar under IFC 98, and the relevant provisions are:

- Clause 2.1 and 2.2 – Possession and completion dates; deferment of possession
- Clause 2.3 to 2.5 – Extension of time
- Clause 2.7 – Liquidated damages for non completion
- Clause 4.11 – Disturbance of regular progress

The employer is again obliged to give you possession of the site on the stated date, but the employer has the right to defer giving possession 'for a time not exceeding the period stated in the Appendix calculated from the Date of Possession, which should not exceed six weeks'. If the employer exercises the power to defer giving you possession you will be entitled to an extension of time and to payment for loss and/or expense. As in the case under JCT 98, the architect also has power to postpone any *work* to be executed under the contract (clause 3.15), and the provisions relating to liquidated damages, extension of time and recovery of direct loss and/or expense are similar to those under JCT 98.

Any action that you take in respect of your rights to extension of time and/or extra cost is commonly known as a 'claim'. The contracts lay down strict procedures which must be followed by both you and the architect. If you do not give the notices etc. which the contract requires at the proper time you may well find that you lose your entitlement under the contract.

These rights are additional and supplementary to your rights under the general law – 'common law claims' – but these can be pursued only by arbitration or court action. The architect has no power to deal with common law claims unless the employer expressly authorises him to deal with them. In some situations, however, you may need to write a suitable letter to the architect or the employer making it clear that you have not waived your common law rights and you should always give very careful consideration to any letter which you may receive from the architect about these matters before you put pen to paper.

It should also be remembered that an adjudicator is only empowered to deal with disputes arising *under* the contract and any relief obtained by an adjudicator's decision will relate to your rights under the contract. It will not extend to any rights you may have under the general law.

5.02 Possession of the site and commencement of work

JCT 98 and WCD 98, both at clause 23.1.1, and IFC 98 at clause 2.1, are quite specific in their terms: 'possession of the site shall be given' to you on the date named in the appendix. Unless JCT 98 and WCD 98, clause 23.1.2 or IFC 98, clause 2.2 apply (which must be stated in the appendix), the employer *must* give you possession of the site on that date and is in breach of contract if he fails to do so. Failure to give possession on the due date is quite a common occurrence and **documents 5.02.1** and **5.02.2** may be appropriate as you ought to reserve your rights.

If, by agreement, you waive your rights to possession on the due date, the terms of the agreement should be recorded in writing and you are in quite a strong bargaining position in these circumstances. If you are approached by the architect and are asked to waive the date for possession, at the very least you should be reimbursed for any costs which you have already incurred and you should certainly not agree to complete within a shorter period than that originally specified. It is our view that these matters should be recorded in a supplemental agreement, signed by both you and the employer, and this should be drafted by a solicitor.

MW 98, clause 2.1, says no more than 'The Works may be commenced on...' (see section 5.12).

If the employer is empowered to defer giving possession of the site and does so this gives you a right to an appropriate extension of time and to reimbursement of any direct loss and/or expense which you suffer or incur as a direct result, and you should write to the architect accordingly: see **document 5.02.3**. Under JCT 98 and WCD 98, clause 23.1.2, if the appendix says that it applies, the employer's right is to defer the giving of possession for a period not exceeding six weeks *or any lesser period stated in the appendix*. Under IFC 98 the employer has the right to defer possession for the period actually stated in the appendix which cannot exceed six weeks unless, of course, clause 2.2 is appropriately amended. **Document 5.02.4** is the sort of letter you should send if any attempt is made to exceed the powers conferred.

Once you have been given possession of the site you must thereupon 'begin the Works' and 'regularly and diligently proceed with' the same and you must complete them 'on or before' the completion date, as extended under the contract. You are entitled to complete the works on a date earlier than the date stated in the appendix but you cannot be compelled to do so even if, as sometimes happens, you have submitted a programme showing intended completion earlier than the contractual date. **Document 5.02.5** is a letter dealing with an architect's attempt to require acceleration to meet an earlier programmed date.

Where you have submitted a programme showing an earlier completion date than the appendix date, there is no obligation to the architect to provide you with drawings and information at times to enable you to meet the programme date, even if the architect approved or accepted your programme without comment. You have no claim in those circumstances.

5.02.1   Letter to the employer about late possession

SPECIAL DELIVERY

To the Employer   *Date*
(Copy to the Architect)

Dear Sir,

[Heading]

Clause [23.1.1 (*JCT 98 or WCD 98*) or 2.1 (*IFC 98*)] of the
Contract between us provides for you to give us possession of
the site on the date specified in the Appendix which is [*date*].
The Architect [Employer's Agent (*WCD 98*)] has today informed
us that you are not in a position to give possession of the site
to us on that date because [*specify reason*].

Clause [23.1.2 (*JCT 98 or WCD 98*) or 2.2 (*IFC 98*)] does not apply
because it is not stated to apply in the Appendix.

As you have no doubt been advised, this amounts to a breach of
Contract by you and we must therefore inform you with regret
that we reserved our common law rights in this matter and in
particular our right to claim compensation for any loss or
damage which we suffer as a result. [*Continue as appropriate*].

Yours faithfully,

5.02.2   Letter to employer about late possession

SPECIAL DELIVERY

To the Employer   *Date*
(Copy to the Architect)

Dear Sir,

[Heading]

Under Clause [23.1.1 (*JCT 98 or WCD 98*) or 2.1 (*IFC 98*)] of the
Contract possession of the site should have been given to us on
[*date*]. Possession of the site was not given to us on that date
and on telephoning the Architect we have been informed that he
is unable to say when we shall be given possession.

Alternative:
[The Architect has informed us today that you will be unable to
give us possession of the site on [*date*] as stipulated by
Clause [23.1.1 (*JCT 98 or WCD 98*) or 2.1 ([*IFC 98*)] because
[*give reason*].]

This is a serious breach of contract since you have no power to
defer giving possession to us on the specified date for
possession in the Appendix and we must inform you with regret
that we reserve our legal rights and remedies in this matter
and in particular our right to claim damages for our resultant
loss.

[Without prejudice, we suggest an urgent meeting in order to
discuss how this difficulty can be overcome.]

Yours faithfully,

5.02.3   Letter when possession deferred

To the Architect (JCT 98, IFC 98)   *Date*
(or Employer under WCD 98)

Dear Sir,

[Heading]

Thank you for your letter of [date] informing us that [the
Employer is (*JCT 98 or IFC 98*) or you are (WCD 98)] exercising
his [your] right under Clause [23.2.2 (*JCT 98 or WCD 98*) or 2.2
(*IFC 98*)] to defer the giving of possession of the site.

In accordance with Clause [25.2.1.1 (*JCT 98*) or 25.2.1
(*WCD 98*) or 2.3 (*IFC 98*)] we hereby give you notice that
progress of the Works is likely to be delayed as a result. As
soon as it is possible for us to assess the effect of the delay
on progress of the Contract we will provide you will full
details when you will no doubt grant us a fair and reasonable
extension of time. Under Clause [26.1 (*JCT 98 or WCD 98*) or 4.11
(*IFC 98*)] we hereby make application for the direct loss and/or
expense which we have incurred [or are likely to incur] due to
this deferment of giving us possession and in support of our
application we enclose details of our loss and expense.
[Enclose vouchered details.]

Yours faithfully,

5.02.4 Letter to employer when possession wrongfully deferred

To the Employer   *Date*
(Copy to the Architect)

Dear Sir,

[*Heading*]

We have received your letter of [*date*] [*or* We have received a letter from the Architect of [*date*]] which purports to defer the giving of possession of the site to us under Clause [23.1.2 (*JCT 98 or WCD 98*) or 2.2 (*IFC 98*)] for [*insert number of weeks*] from [*insert date for possession as stated in Appendix*].

We regret to inform you that this action is invalid because [*give reason, e.g.* (under IFC 98), Clause 2.2 only permits you to defer the giving of possession for a time not exceeding the period stated in the Appendix, which is three weeks, and your letter purports to defer possession for four weeks; *or* (under JCT 98 or WCD 98) Clause 23.1.2 only entitles you to defer the giving of possession for a period not exceeding six weeks, and you are purporting to defer possession for two months (*or* an indefinite period)]. We cannot therefore accept your action and ask that your/his letter be withdrawn.

If you insist on deferring possession in this way, this constitutes a breach of contract, in respect of which we reserve our rights and remedies at common law, and in particular our right to claim damages for breach. Without prejudice, however, we are prepared to meet you to discuss how the problem might best be overcome.

Yours faithfully,

5.02.5   Letter to architect where required to complete by programme date

---

To the Architect   *Date*

Dear Sir,

[*Heading*]

We have received your letter of [*date*] in which you state that
unless we complete by [*insert date*] we shall be in breach of
contract and be liable to liquidated and ascertained damages.
Your letter also instructs us to increase our labour resources
so as to enable completion by that date which was the date shown
in the programme we submitted to you on [*insert date*].

We would respectfully point out to you that the Contract
Appendix gives the Date for Completion as [*insert date*] and our
contractual obligation is to complete 'on or before' that
date. The programme date has no contractual significance and
we would only be in breach of contract if we failed to complete
by the contract Date for Completion. You have no power under
the Contract to order acceleration of the Works and we very
much regret that we must ignore your letter. Unless we fail to
complete the Works by [*insert contractual date*] any purported
deduction of liquidated and ascertained damages by the
employer will be invalid.

[If early completion is critical to the Employer, we will be
happy to discuss with you how this might be achieved by the use
of extra resources provided that the employer is prepared to
agree to pay for any acceleration measures which are agreed.]

Yours faithfully,

5.03   Delay in general

'To delay' means to slow up or cause to be late. Both clause 25 of JCT 98 and WCD 98 and clause 2.3 of IFC 98 require you to notify the architect of *any* delay and not merely delay which may entitle you to an extension of time. You are only entitled to an extension of time for delay caused by those events listed in clause 25.4 and 2.4 respectively but the architect is entitled to take notice of *any* delay which is affecting or is likely to affect progress. Clause 2.2 of MW 98 only requires this notice if the delay is for reasons beyond your control.

Document 5.03.1 is a letter notifying the architect of a cause of delay which is not ground for extension of time. **Document 5.03.2** is an example of a letter notifying the occurrence of an event which gives rise to a claim for extension of time.

Some contractors conveniently overlook the fundamental obligation placed on them by JCT 98 or WCD 98, clause 25.3.4, and the proviso to IFC 98, clause 2.3, to use constantly their best endeavours to prevent delay and to do all that may *reasonably* be required to the satisfaction of the architect (employer WCD 98) to proceed with the works. Conversely, some architects assume that this amounts to an obligation to mitigate a delay which has already occurred or which is unavoidable. Where the delay is caused by an event which is a ground for extension of time this is not so nor can the architect require you to spend additional money, for instance by accelerating, to meet this obligation. Neither JCT 98 nor IFC 98 empowers the architect (nor WCD 98 the employer) to instruct you to accelerate and **document 5.03.3** is a letter in response to such an architect.

5.03.1   Notification of delay by an event which is not a ground for extension of time

---

To the Architect   *Date*
(or *Employer under WCD 98*)

Dear Sir,

[*Heading*]

We regret to inform you that progress of the Works is currently being, and is likely to continue to be, delayed by a breakdown in the tower crane. Since this is of Japanese manufacture and the local agents are unable to effect repairs, technicians are being flown in from Japan and it will be several days at least before it is again usable.

Needless to say, we are making every effort to overcome this problem in the meantime and we shall be taking steps to ensure that completion of the Works is not delayed.

We shall inform you as soon as the problem is overcome and, in the meanwhile, please accept this as the formal notice required by Clause [25.2.1.1 (*JCT 98*) or 25.2.1 (*WCD 98*) or 2.3 (*IFC 98*)] of the Contract.

Yours faithfully,

5.03.2 Notification of delay caused by an event which is a ground for extension of time

---

To the Architect   *Date*

Dear Sir,

[Heading]

In accordance with Clause [25.2.1.1 (*JCT 98*), 25.2.1 (*WCD 98*),
2.3 (*IFC 98*) or 2.2 (*MW 98*)] of the Contract we hereby give you
notice that progress of the Works is being and is likely to
continue to be delayed by continuous snow and frost conditions
which are effectively preventing work from continuing on
site. As you are aware, this work currently consists of
excavation and concreting for the foundations of Block 2 and it
is not possible for us, at this stage in the Contract, to find
alternative work on site.

These circumstances amount to 'exceptionally adverse weather
conditions' which is [a relevant event under Clause 25.4.2
(*JCT 98 or WCD 98*) *or* an event qualifying as a ground for
extension of time under Clause 2.3 (*IFC 98*) *or* a reason beyond
our control (*MW 98*)]. Although we programmed to allow for
normal weather conditions, it is our view that heavy snow and
continuous frost are exceptional in this locality in late
April.

Since these conditions are continuing and the long-range
weather forecast shows them as likely to continue for some
time, [(*JCT 98 or WCD 98*) we are unable at this stage to give the
particulars and estimate required by Clause 25.2.2 of the
Contract. We shall, of course, provide you with these
particulars and estimate as soon as possible. In the meantime,
we are using and will continue to use our best endeavours to
prevent delay, but as we have already indicated there is very
little that we can do at the present time.

When we have provided the particulars and estimate required,
no doubt you will grant the appropriate extension of time
within the time limit laid down in Clause 25.3.1.]

(cont.)

(5.03.2   cont)

[*or (IFC 98 or MW 98)* : . . . we cannot at this stage provide you
with any further information as to the expected effects of the
delay but we will do so as soon as possible. In the meantime, we
confirm that we are using our best endeavours to prevent
delay.]

Yours faithfully,

5.03.3   Letter where the architect requires acceleration etc.

To the Architect   *Date*

Dear Sir,

[*Heading*]

Further to our notice of delay dated [*date*] and your letter of [*date*] in reply, we note that your letter purports to instruct us to take steps, following cessation of the current exceptionally adverse weather conditions, to accelerate so as to make up lost time.

We would respectfully point out to you that the Contract does not empower you so to instruct us and we are not prepared to take any such steps except with the express agreement of the Employer to reimburse us for the additional costs involved plus a reasonable allowance for our overheads and profit. Such an agreement would, of course, be supplementary to the Contract and its terms will have to be negotiated with the Employer to whom we are sending a copy of this letter.

[*If appropriate:* Contrary to the view expressed in your letter [Clause 25.3.4 (*JCT 98 or WCD 98*) *or* the proviso to Clause 2.3 (*IFC 98*)] does not empower you to instruct acceleration and you have no inherent power to do so.]

Yours faithfully,

5.04   Detailed procedures under JCT 98 or WCD 98

You are not merely required to notify delay; as **document 5.03.2** shows you are also required to provide *particulars* of the expected effects of the relevant event you have identified and give your own estimate of any expected delay to the completion of the works resulting therefrom. Even if your notice refers to more than one relevant event, you must give the particulars and estimate relating to each such event as if it were the only one.

If possible, you must give the particulars and estimate in the notice; if not, you must give them as soon as possible afterwards. Any undue delay in providing the particulars and estimate might result in your losing your entitlement in certain circumstances.

The particulars which you provide must be as full and detailed as possible so that the architect (employer WCD 98), sitting in his office, can form his own judgment as to the merits of your case and the extension of time which you will genuinely need. Do not overstate your case as this could antagonise the architect or employer and lead to the mistrust which is at the root of so many contract problems.

Document 5.04.1 is a letter which is both the required notice and the particulars and estimate. **Document 5.04.2** gives particulars and estimate after notice has been issued.

On many large-scale projects under JCT 98 nominated subcontractors will be involved. Where your notice includes reference to a nominated subcontractor, you must send a copy of that notice to him immediately. The contract refers to 'the material circumstances including the cause(s) of delay' and it is quite clear that a nominated subcontractor is entitled to a copy of the notice, not only if he is the cause or a cause of the delay but also where the delay will affect the progress of his work, whether he has commenced it or not. His subcontract gives him rights and obligations with regard to delay similar to yours and he must be given the opportunity to give notice of delay to his work to you. The nominated subcontractor may not be as aware of the circumstances as you are.

5.04.1   Letter giving notice, particulars and estimate

To the Architect (JCT 98)   *Date*
(or Employer under WCD 98)

Dear Sir,

[*Heading*]

We hereby give you notice under Clause 25.2.1.1 (*JCT 98*)
2.5.2.1 (*WCD 98*) of the Contract that compliance with your
Instruction No. 173 of [*date*] received today is likely to cause
delay to the progress of the Works. Compliance with this
instruction is a relevant event under Clause 25.4.5.1.

Particulars of the expected effects of compliance are that the
additional work will take a further three weeks to complete and
no other work will be possible while this is being carried out.
As you will see from the attached copy of our programme, on
which we have indicated in red the work required by your
instructions, we are at a critical stage. The intervention of
this work will also affect the work of subsequent trades which
will lead to still further delay in the execution of that work
as we have indicated in blue on the enclosed programme.

In these circumstances we estimate that the likely delay to
Completion of the Works beyond the Current Completion Date
will be five weeks. Please confirm that the particulars and
estimate given above are sufficient to enable you to grant the
appropriate extension of time as required by Clause 25.3 of the
Contract. If you require any further particulars please
specify their nature within the next 7 days, failing which we
will assume that the information we have given is sufficient.

[(JCT 98) Clearly, the work of all nominated subcontractors
will also be affected and we are therefore sending a copy of
this notice to each of them as required by Clause 25.2.1.2 of
the Contract.]

Yours faithfully,

5.04.2   Particulars and estimate given after notice

To the Architect (JCT 98)   *Date*
(or Employer under WCD 98)

Dear Sir,

[*Heading*]

Further to our notice of delay of [*date*] we now give below the
particulars and estimates relating to the relevant events
identified therein as follows:

1. Civil commotion (Clause 25.4.4)
The police have now removed the nuclear protesters who have
been occupying the site and since they are in custody now no
similar problems are anticipated in the immediate future.
During their occupation of the site work totally ceased for a
period of five working days. It has taken a further two days to
remove the debris and repair the damage to the perimeter fence
and we have taken further steps to ensure the security of the
site. We therefore estimate that the delay in completion of the
Works before the current Completion Date will be 7 working days
under this head.

2. Late instructions (Clause 25.4.6)
We have now received from you the joinery details which we
specifically requested you to provide in our application of
[*date*]. The date on which it was necessary for us to receive
these details, having regard to the Completion Date currently
fixed, was [*date*]. The delay in receipt of these details was
therefore 4 weeks. However, while affecting progress of the
Works, it was not an absolutely critical factor, and we
therefore estimate the delay in completion of the Works beyond
the current Completion Date resulting from this will be 12
working days. *Continue as appropriate.*]

                                                    (*cont*)

(5.04.2   cont)

As required by Clause 25.2.2 of the Contract these particulars
and estimates are given for each relevant event considered
without regard to other relevant events. No doubt you will
consolidate these into a single grant of extension of time
after taking all factors into account.

We look forward to receiving your grant of an extension of time
stating the new Completion Date within the period specified in
Clause 25.3.1.

Yours faithfully,

5.05 The grant of the extension

Once you have provided the architect or employer with reasonable particulars and your estimate of the delay, he is bound to grant any appropriate extension of time within 12 weeks, stating the new completion date. If there are less than 12 weeks to the completion date currently fixed he must grant the extension no later than that date. If the architect or employer decides not to fix a new completion date he must notify you in writing of his decision, and the time limit of 12 weeks applies.

However, you are required to keep the architect or employer up to date by providing further notices, particulars and estimates as may be necessary in relation to previous notices of delay (clause 25.2.3 JCT 98 or WCD 98) (**document 5.05.1**). Moreover, the architect or employer may consider that he needs more information before he can form a proper judgment. If the information which he requests is reasonably necessary you must provide it (**document 5.05.2**) and his obligation to grant an extension within 12 weeks commences on receipt of this further information – provided, of course, that it itself is reasonably sufficient and that he has no good reason to ask you for more.

You may disagree with the extension when granted, but remember that the architect or employer is only bound to grant an extension of time 'by fixing such later date ... as he *then* estimates to be fair and reasonable'. Such extended date is the date at which you must aim. Your protection is the architect's duty to review his previous decision within 12 weeks of practical completion of the works, having regard to all the circumstances of which he is then aware.

If it is reasonable for you not to have completed by the completion date which the architect/employer has previously fixed he should grant you the necessary further extension at this time. In any case, the architect's/employer's decision is subject to review. This review could be started immediately by issuing a notice of adjudication or left until some later date when the matter can be referred either to adjudication or to arbitration. If you consider that any decision of his is unreasonable you would do well to register your protest (**document 5.05.3**) and, if you feel sufficiently aggrieved by the decision, even give notice of adjudication immediately.

Where there are fewer than 12 weeks before the current completion date, the architect/employer must make his decision no later than that date. If he fails to do so, you may well possibly argue that 'time is at large' and if that contention is correct you would no longer be subject to liquidated damages. You would certainly need to take legal advice at this stage.

Many architects/employers are slow in giving extensions of time, but the contract is quite clear and the architect/employer must strictly observe the time limits. If the completion date is reached before practical completion occurs – which is not uncommon – the architect/employer may confirm the existing date or change it. Once the works are practically complete the architect/employer is in a difficult position if he fails to carry out the duty imposed on him under clause 25.3.3. He has no choice; he must write to you not later than the expiry of 12 weeks from the actual date of practical completion and do one of three things:

- Confirm the current completion date.
- Grant a further extension if circumstances justify it.
- Fix an earlier date – but this can be no earlier than the original date for completion specified in the appendix. This option is only open to him if, since the completion date was last fixed, he has issued instructions omitting work which, in his opinion, reduced the time needed to complete the remaining work.

The sanction against an architect/employer who fails to take action under clause 25.3.3 within the 12 week period is to issue immediate notice of adjudication especially if the employer gives notice of his intention to impose liquidated damages. Architects often delay carrying out the duty until after the 12 week time limit. This is bad contract practice and the only counter is an immediate notice of adjudication.

In granting extensions the architect/employer can take into account any omission of work which he has instructed since the completion date was last *fixed*. It is plain from the wording of the contract that he may do this in his first grant of an extension provided that he has omitted work before the grant and the only limitation is that you are always entitled to the contract period as originally specified and at no time can the architect fix a completion date earlier than the original date for completion stated in the appendix to the contract. **Document 5.05.4** is the sort of letter you should write if the architect/employer in fact exceeds his powers.

5.05.1 Supplementary notice, particulars and estimate

To the Architect (JCT 98)   *Date*
(or Employer under WCD 98)

Dear Sir,

[Heading]

Further to our notice of delay of [*date*] and the particulars
and estimates provided therein submitted on [*date*] as
required by Clause 25.2.3 of the Contract we now give you
further notice that the progress of the Works is being further
delayed by the circumstances notified.

*[Set out as appropriate any material changes in the
circumstances, particulars and estimate, e.g.* (*JCT 98*) Messrs
XYZ Ltd, the nominated subcontractor for the heating
installation, have now notified us that, owing to personnel
problems, they will not be able to resume Works for at least two
weeks. We attach a copy of their notification and notice under
NSC/C Clause 2.2. We are sending a copy of this letter to them.

(*WCD 98*) We have now ascertained that the damage caused by the
nuclear protesters included removal by them of certain
amounts of bar reinforcement which we have been unable to
recover and it will take 10 days to obtain replacement
materials.]

*[Detailed particulars of further delay and any revision of
estimate.]*

Yours faithfully,

5.05.2   Letter where further particulars, etc. are requested

To the Architect (JCT 98)   *Date*
(or Employer under WCD 98)

Dear Sir,

[Heading]

Thank you for your letter of *[date]* requesting further
particulars under Clause 25 of the Contract.

*[Set out required information following terms of Architect's/
Employer's letter]*
*[Or where Architect requests further notice under Clause
25.2.3 to keep particulars and estimate up to date, give him
the supplementary information. For example:*
We have now been able to ascertain that by working around the
problem areas we have been able to contain the overall delay to
progress caused by the need to obtain replacement bar
reinforcement to no more than 5 days.*]*

Yours faithfully,

5.05.3   Letter protesting at the extension granted (JCT 98)

To the Architect   *Date*

Dear Sir,

[*Heading*]

We acknowledge receipt of your grant of extension of time dated
[*date*] from which we note the Completion Date which you have
now fixed.

We draw your attention to our notice of delay of [*date*] in which
we estimated the expected delay in completion of the Works
beyond the Completion Date then fixed as [*x*] weeks, which would
have given a new Completion Date of [*date*].

With respect, in our view this estimate was a reasonable one in
the light of the circumstances and particulars which we
provided. We do not understand your decision to reduce our
estimate so substantially and we must record the fact that we
cannot accept that the Completion Date you have now fixed is
fair and reasonable.

We must ask you to reconsider your decision as a matter of
urgency, failing which we shall be compelled to give notice of
adjudication.

Yours faithfully,

5.05.4 Letter where the architect exceeds his powers regarding the fixing of an earlier completion date (JCT 98)

SPECIAL DELIVERY

To the Architect   *Date*
(Copy to the Employer)

Dear Sir,

[Heading]

We have received what purports to be your grant of an extension of time under Clause 25.3 in which you state that, pursuant to Clause 25.3.1.4, you have taken into account the omission of work instructed by your AI No. [*specify number*] of [*date*].

We must point out that this instruction was issued before your last grant of an extension of time on [*date*]. This omission must therefore be deemed to have been taken into account in that extension and cannot be taken into account in the extension which you have currently granted.

We shall be pleased if you will rectify this matter immediately. [Should you fail to do this we shall be forced to seek an adjudicator's decision on this matter.]

Yours faithfully,

5.06   Procedures under IFC 98

The procedures for the grant of extensions of time under clause 2.3 of IFC 98 are essentially a scaled-down version of JCT 98, clause 25, but there are some significant differences. Thus, although you are not required to provide particulars of the expected effects of delays and your own estimate of the resulting delay in completion, you are required to provide such information required by the architect/contract administrator 'reasonably necessary' to enable him to grant a proper extension of time. You must, therefore, provide the architect/contract administrator with the information he needs and it is in your interest to provide that information as soon as possible. **Document 5.06.1** is a suggested notice of a cause of delay which is a ground for extension of time. The architect/contract administrator may reasonably require you to provide further information and **document 5.06.2** is a suggested reply to such a request.

Unfortunately, there is no specific time limit laid down within which the architect/contract administrator must deal with extensions of time after your notice. The contract merely says that the extension is to be granted 'so soon as he is able to estimate the length of the delay' which is delightfully vague. However, the case law suggests that you must be informed of your new completion date as soon as reasonably practicable. If the sole cause of the delay is, for example, a variation ordered under clause 3.6, the effect of which is immediately apparent, then the architect/contract administrator ought to grant an extension of time almost immediately so that you have a target for which to aim. Failure to grant the extension within a reasonable time may then invalidate the liquidated damages clause. In contrast, if the cause of the delay is outside the control of the employer or the architect/contract administrator, i.e. is a neutral event, especially where its duration is uncertain, then the grant of an extension may be delayed, but even here we think that the extension should be granted shortly after the factors which govern the exercise of the architect's/contract administrator's discretion have been established. Where there are multiple causes of delay extending over a considerable period, then the architect/contract administrator may have no alternative but to leave the matter until late in the day, i.e. until the 12 week period after practical completion. **Document 5.06.3** is a letter to a dilatory architect/contract administrator. This example does not mention adjudication but you may feel that the threat of a notice of adjudication is appropriate in certain circumstances.

The architect/contract administrator is not bound to inform you if he decides not to grant an extension of time, though it would be discourteous of him not to do so. A few architects/contract administrators consider that IFC 98 empowers them to reduce extensions which they have granted previously where they have subsequently omitted work. This is not so and under IFC 98 the architect/contract administrator cannot withdraw or reduce an extension of time already granted and if the architect/contractor administrator purports to do so you should write to him at once – **document 5.06.4**.

Under IFC 98 the architect/contract administrator is not *obliged* to review the extensions of time granted; the contract merely empowers him to grant an extension of time or review a previous decision if he wishes at any time up to 12 weeks after the date of practical completion, and he can do this whether or not you have given him notice of delay. The purpose of this provision is to preserve the employer's right to liquidated damages because, if completion is delayed because of an event which is the employer's responsibility and no extension of time is awarded, the contract time will be 'at large' and the employer will lose his right to liquidated damages. This will not necessarily let you off the hook because the employer can always sue for common law damages if he can prove that he has suffered loss. If you consider that time has become 'at large' in this way you should seek immediate legal advice and a letter along the lines of **document 5.06.5** may be indicated.

In this instance as the dispute relates in part to time being 'at large' arbitration may be a better forum than adjudication in order to avoid possible problems arising from arguments that such a matter does not arise 'under' the contract. **Document 5.06.5** reflects this suggestion.

Other letters in section 5.05 may be adapted as appropriate if needed.

5.06.1 Notification of delay caused by an event which is a ground for extension of time (IFC 98)

To the Architect/Contract Administrator   *Date*

Dear Sir,

[*Heading*]

In accordance with Clause 2.3 of the Contract we hereby give
you notice that the progress of the Works is being and is likely
to be delayed by the failure of Southshire Electricity to
commence work on the new electric main and substation. This is
effectively preventing us from doing any other work on site and
in our view is a qualifying event under Clause 2.4.13. We
understand from Mr Idle, Southshire Electricity's Regional
Manager, that the work cannot start until six weeks from today.

We enclose (a) a letter from Mr Idle confirming delayed start;
(b) a copy of our revised programme which reflects work
actually done on site to today's date from which you will see
that the electricity company's work is on the critical path;
(c) a programme projected on the basis that the electricity
company will commence work on the new date and assuming that
they complete their work on time; and (d) a letter which we have
sent to [*a named subcontractor; subcontract work will be
delayed*]. You will see that we have asked him to let us have his
revised programme and to send a copy to you.

We hope that this information is sufficient to enable you to
grant the necessary extension of time, but if you require any
further information you will no doubt let us know.

Yours faithfully,

5.06.2 Letter to architect/contract administrator providing further information (IFC 98)

To the Architect/Contract Administrator   *Date*

Dear Sir,

[*Heading*]

Thank you for your letter of [*date*] asking us to provide
further information about the run-on effect of the current
delay.

The further information you require is:

[*Give information as reasonably requested by the architect
supported by any necessary documentation.*]

We hope that this additional information will now enable you to
grant the extension of time in the near future.

Yours faithfully,

5.06.3   Letter to architect/contract administrator who has failed to grant extension of time (IFC 98)

SPECIAL DELIVERY

To the Architect/Contract Administrator   Date
(Copy to the Employer)

Dear Sir,

[Heading]

On *[date]* we sent to you a formal notice of delay under Clause
2.3 of the Contract as a result of your AI No. *[insert number of Instruction]* whereby you varied the work by requiring us to
[specify]. We gave you the further information you requested
on *[insert date]* and you confirmed that it was necessary to
carry out the variation before any further work could proceed.

[X] weeks have now elapsed since this date and we have heard
nothing from you. Will you please grant the appropriate
extension of time immediately in fairness to both ourselves
and to the employer to whom we are sending a copy of this letter
as we are advised that your failure to grant an extension of
time within a reasonable time of the issue of the variation
order may lead to the contract time being 'at large' with the
consequent forfeiture of the employer's right to deduct
liquidated damages should we not complete on time. By any
stretch of the imagination, you have had more than a
'reasonable time' in which to make your decision.

Yours faithfully,

5.06.4 Letter to architect/contract administrator who wrongfully reduces extension of time (IFC 98)

To the Architect/Contract Administrator   *Date*

Dear Sir,

[*Heading*]

We have received your letter of [*date*] and note with some surprise that by it your purport to reduce the extension of time which you granted on [*insert date*] and give us 'a revised completion date of [*date*]' on the basis that you have subsequently omitted [*specify omission*].

As you are aware, this Contract is on IFC 98 terms, and Clause 2.3 does not empower you to withdrawn or reduce extension of time already granted even if you have omitted work in the interim. Consequently, we regard your letter as a nullity and shall be glad to have your written confirmation that the Date for Completion is that previously fixed by you, namely [*insert date*].

Yours faithfully,

5.06.5   Letter where time is 'at large' (IFC 98)

SPECIAL DELIVERY

To the Architect/Contract Administrator   *Date*
(*Copy to the Employer*)

Dear Sir,

[*Heading*]

On [*date*] we gave you notice of delay under Clause 2.3 of the
Contract because of the Employer's deferment of possession of
the site for six weeks, which is an event giving rise to an
extension of time [Clause 2.4.14]. Despite our letter of
[*date*] reminding you that an extension of time was due and
requesting an immediate decision, we heard nothing from you.

Practical Completion of the Works was certified by you as being
achieved on [*date*]. Your decision on extensions of time under
the Contract should have been made no later than 12 weeks from
that date, i.e. by [*date*].

Since you have made no such decision, and no longer have power
to do so, we are advised that the Employer is no longer entitled
to recover or deduct liquidated damages because the time for
completion has become 'at large'.

We have decided to refer this matter to arbitration and enclose
our Notice to the Employer accordingly.

Yours faithfully,

5.07   Liquidated and ascertained damages

A number of contractors are notorious for alleging that the liquidated damages figures applied in building contracts are extortionate. In our experience the converse is the case and you may well benefit from what is, in fact, a limitation on your liability in respect of damages for non-completion. The important point about liquidated damages is that they are recoverable without proof of loss. Even if you were successful in establishing that a particular figure amounted to a 'penalty' and was irrecoverable, the employer would be able to recover unliquidated damages against you for his proven losses arising from late completion.

The employer does not have to justify the amount of liquidated damages to you before he applies the clause. What he does have to do, under JCT 98, WCD 98 and IFC 98, is to notify you in writing of his intention to recover them, in whole or part, and he must do this before making any deduction from any payment due to you and in any case no later than the date of the final certificate. (This is a withholding notice as required by the contract conditions.) The architect must first issue a certificate of non-completion under clause 24 (JCT 98) or clause 2.6 (IFC 98); the employer does this himself by a notice under clause 24 of WCD 98. If, having done so, a further extension of time is granted and a new completion date is fixed, and you have not completed by that date, a new certificate (JCT 98) or notice (WCD 98) must be issued and the employer must give you a fresh notice of his intention that he requires liquidated damages since once a new completion date is fixed any notice given by the employer before it is rendered void.

Upon a subsequent variation to the completion date the provisions are that the employer repays any amounts recovered, allowed or paid based on the completion date that no longer applies.

The agreed figure is recoverable even if, in the event, the employer suffers substantially less or even no loss. Nevertheless the employer's entitlement to liquidated damages is dependent on the architect having carried out his obligations properly as to extension of time in accordance with the terms of JCT 98 clause 25 or IFC 98 clause 2.3 as appropriate. The employer is also disentitled from recovering liquidated damages if he (or those for whom he is responsible in law) are in any way responsible for delaying completion of the works in ways for which an extension of time cannot be granted, e.g. by failing in good time to give the necessary consent to statutory undertakers regarding the laying of new mains etc. In such a case you should write to the employer appropriately (**document 5.07.1**) and also seek legal advice.

Although there is a sectional completion supplement to both JCT 98 and IFC 98 designed for a specific purpose, many employers (and particularly local authority employers) fail to use it when they require phased completion, e.g. when on a housing contract they intend to take possession of each block of dwellings as it is completed. One completion date is inserted in the appendix and often the liquidated damages are expressed as '£x per week per dwelling' or '£x per week per block of dwellings'.

This raises problems both as regards the granting of extensions of time and the enforcement of liquidated damages. Case law has thrown doubt on the basic validity of such an arrangement and it seems that such provisions may be unenforceable, unless the sectional completion supplement is used or the contract form is substantially amended. Provision in the bills or other contract documents for phased completion is not sufficient.

If you are faced with this sort of situation, it is worth seeking legal advice, and **document 5.07.2** may be used, amended according to the particular facts.

5.07.1   Letter challenging the employer's right to deduct liquidated damages

SPECIAL DELIVERY

To the Employer   Date
(*Copy to the Architect*)

Dear Sir,

[*Heading*]

We note that you have deducted the sum of £[*x*], being
liquidated damages at the rate of £[*y*] per week for [*specify
number*] weeks, from the sum certified as due to us in
Certificate No. [*specify number*].

You are not entitled to make such a deduction because:
1.   you have failed to notify us in writing of your intention to
do so as required by Clause [24.2.1 (*JCT 98 or WCD 98*) or 2.7
(*IFC 98*)] of the Contract, *or*
2.   the Architect has not yet properly extended time under
Clause [25 (*JCT 98 or WCD 98*) or 2.3 (*IFC 98*)] of the Contract
[*or as appropriate*], *or*
3.   the Architect has not yet issued the necessary Certificate
of delay under Clause [24.1 (*JCT 98 or WCD 98*) or 2.6 (*IFC 98*)]
of the Contract, *or*
4.   through your failure timeously to give your consent to
Southshire Electricity in respect of the laying of new mains
serving the site, you are responsible in whole or in part for
the alleged delay, *or*
5.   you have failed to issue the written notice required by
Clause [30.1.1.4 (*JCT 98*) or 30.3.4 (*WCD 98*) or 4.2.3(b) (*IFC
98*)]

Unless we receive your cheque for £[*x*], being the amount
wrongfully deducted, within 7 days from today's date we shall
issue a Notice of Adjudication under Clause [41A.4.1 (*JCT 98*)
or 39A.4.1 (*WCD 98*) or 9A.4.1 (*IFC 98*)] and/or exercise the
right to determine our employment under Clause [28 (*JCT 98 or
WCD 98*) or 7.5 (*IFC 98*)].

Yours faithfully,

5.07.2 Letter challenging the validity of liquidated damages for phased completion where there is no sectional completion supplement and the contract is unamended

To the Employer   *Date*
(Copy to the Architect)

Dear Sir,

[Heading]

We acknowledge receipt of your letter of [*date*] informing us of your intention to deduct liquidated damages from monies due to us in respect of those dwellings remaining uncompleted at the Completion Date currently fixed at the weekly rate of £50 per week for each uncompleted dwelling.

Since no sectional completion supplement has been entered into, and the necessary amendments have not been made to the printed Contract, we are advised that your purported deduction is invalid.

In the circumstances, no doubt you will take appropriate advice before making any deductions, and unless we receive the full sums certified, without deduction for liquidated damages, we shall be obliged to issue a Notice of Adjudication under Clause [41A.4.1 (*JCT 98*) or 39A.4.1 (*WCD 98*) or 9A.4.1 (*IFC 98*)]. Furthermore, we reserve our right to take appropriate action under Clause [28 (*JCT 98 or WCD 98*) or 7.5 (*IFC 98*)] of the Contract between us.

Yours faithfully,

5.08   Reimbursement of loss and expense under JCT 98 and WCD 98

Your right to reimbursement of direct loss and/or expense is set out in clause 26 of both JCT 98 and WCD 98, which lays down a procedure which must be followed. The machinery is relatively simple but if you fail to operate it correctly you may well lose your right to reimbursement under the contract by way of certification by the architect (JCT 98) leaving you to pursue any entitlement which you may still have by the ordinary processes of law.

It is not every loss or expense which is recoverable under this clause, nor indeed is every loss recoverable at law. The limiting factor is the word 'direct' which means that the loss or expense claimed must be a direct result of the specified cause. It is not sufficient to say that the loss would not have arisen had the cause not occurred; you must be able to say the loss arose *because* of the cause and nothing happened in between the cause and its effect.

The specified causes are:

- Deferment of giving possession of the site (if clause 23.1.2 applies).
- Not receiving in due time necessary instructions etc. from the architect for which you specifically applied in writing at the correct time. This includes instructions for expenditure of provisional sums (including failure to comply with information release schedule under JCT 98).
- Opening up for inspection or testing of any work or materials unless the inspection or test showed that they were not in accordance with the contract.
- Delay in receipt of permission or approval for the purposes of development control requirements (WCD 98 only).
- Discrepancies in or divergences between the contract drawings and/or the contract bills (JCT 98 only).
- Execution of work not forming part of the contract by the employer or someone else employed or engaged by him, or failure to execute such work, or the supply by the employer of materials or goods which he has agreed to provide for the work, or his failure to supply them.
- Postponement of any work by the architect (JCT 98) or employer (WCD 98).
- Failure by the employer to give access to the site over any land etc. in his own possession and control.
- Variations (changes WCD 98) or work against provisional sums ordered by the architect (JCD 98) or employer (WCD 98).
- Inaccurate approximate quantities. Where the bills include an approximate quantity for work and it is not a reasonably accurate forecast of the work actually done (JCT 98 only).
- Compliance or non-compliance by the employer with clause 6A.1 relating to planning supervision.
- Suspension by the contractor of the performance of his obligations pursuant to clause 30.1.4.

If you suffer or incur loss or expense from any other cause, e.g. because the employer has interfered with your work, you cannot recover it under this clause. You must proceed under common law. Further, the direct loss and/or expense must not be recoverable under any other contract provision, e.g. the effects of inflation recoverable by the operation of the fluctuations clause.

5.09 Machinery for recovery

The first step is for you to make a written application to the architect (JCT 98)/ employer (WCD 98) stating that you have incurred or are likely to incur direct loss and/or expense because regular progress of the works, or any part of them, has been or is likely to be materially affected by one or more of the specified causes. This application must be made as soon as it has become, or should reasonably have become, apparent to you that regular progress has been, or is likely to be, affected. The application must therefore be made at the earliest possible time if it is not to be rejected by the architect/employer. Early notice is required – if reasonably possible before the event – so that the architect may have an opportunity of taking remedial action. **Document 5.09.1** is a form of written application.

The architect/employer may and probably will require you to provide further information so that he may form an opinion as to whether your application if justified. This information should be as detailed as possible and backed up by supporting documentation, e.g. programmes. **Document 5.09.2** is a sample response to an architect's/employer's request.

The architect/employer (or the quantity surveyor (JCT 98)) may and probably will require you to provide details of your loss and/or expense as and when you incur it. Figures should be factual and again as much detail as possible should be given, supported by documentary evidence, e.g. invoices, time sheets etc. **Document 5.09.3** suggests a form which such details might take.

Entitlements to recover loss and/or expense are quite separate from entitlements to extension of contract time. Merely because the contract period is extended, no right to money is conferred. Conversely, there can be a money claim under clause 26 where the contract period is not extended at all. Your entitlement to money will arise from whether, and the extent to which, your work is in fact delayed or disrupted, whereas extensions of time are essentially estimates of delay. As far as JCT 98 is concerned, clause 26.3 unfortunately appears to suggest a connection between extensions of time and money entitlement which leads some architects to argue that you cannot claim direct loss and/or expense unless the contract period is extended and then only to the extent that it is extended. You should resist such attempts to limit your entitlement (see **document 5.09.4**).

Clause 26.3 (JCT 98) requires the architect to state to you in writing what extension of time he has granted in respect of the events for which you are claiming reimbursement. While we deplore the inclusion of this clause, nevertheless it is there, and you are entitled to take advantage of it. **Document 5.09.5** suggests how you might do this.

The whole of clause 26 supposes that claims will be made and met periodically during the course of the contract. You must ensure that the machinery is properly operated in your favour. Do not leave 'claims' until the end of the day. Similarly, you should resist those architects and quantity surveyors (employers under WCD 98) who seek to postpone proper consideration of your claims. **Document 5.09.6** is a potential line of approach in this situation.

If you are dissatisfied with amounts of direct loss and/or expense ascertained and certified, you should register your dissatisfaction immediately (see **document 5.09.7**). If this has no result you should initially pursue your claim in adjudication: see Chapter 9.

In these days of adjudication it is far preferable to get things out into the open as early as possible. The consideration of a detailed application for the costs of delay and disruption by an adjudicator at an early stage may well mean the avoidance of a lengthy argument later on.

5.09.1 Written application for reimbursement of direct loss and/or expense

---

To the Architect (JCT 98)   *Date*
(or Employer under (WCD 98))

Dear Sir,

[*Heading*]

In accordance with Clause [26.1 (*JCT 98 or WCD 98*) or 4.11 (*IFC 98*)] of the Contract we hereby make application to you that we are likely to incur direct loss and/or expense in the execution of this Contract, for which we will not be reimbursed by a payment under any other provision of the Contract because [the regular progress of the Works is likely to be materially affected by [e.g. compliance to your instruction No. [*specify number*] of [*date*] which requires a variation of the work].

[*For adaptation for IFC 98 see section 5.11 later*]

In support of our application we submit the following information and details: [*detail*].

Yours faithfully,

5.09.2 Response to the architect's request for further information

To the Architect (JCT 98)   Date
(or Employer (WCD 98))

Dear Sir,

[*Heading*]

Thank you for your letter of [*date*] requesting us to provide
further information in respect of our application of [*date*].

As you will see from the enclosed programme, compliance with
your AI No. [*specify number*] which requires us [to make
substantial alterations to the roof structure to accommodate
additional dormer windows (*JCT 98*) or redesign and make
substantial alterations to the roof structure (*WCD 98*)] will
materially affect the progress of the work in that it
necessitates the ordering of additional timbers which, as we
have already commenced the roof structure as originally
designed, will inevitably mean substantial delay and
disruption to our working. We shall be unable to find other
work on the site for our carpenters while [this additional
timber is being obtained (*JCT 98*) or design work is procured
and additional materials obtained (*WCD 98*)] and we must
therefore deploy them elsewhere so as to mitigate the
financial effect. Even when the additional materials are
delivered, it will be some time before we can muster a full
force of carpenters again.

The delays in the roof structure will mean that the roof
finishing will be delayed, and this will lead to a delay in the
internal finishing trades which cannot be commenced until the
building is weather-tight.

We are notifying our subcontractors of this delay, but we
obviously cannot expect them to be available at short notice
and therefore there may well be further delays and possible
claims from them.

We shall, of course, keep you fully informed.

Yours faithfully,

5.09.3   Details of loss and/or expense

To the Quantity Surveyor   *Date*
(Copy to the Architect) (JCT 98)
(or Employer under (WCD 98))

Dear Sir,

[Heading]

In response to your letter of *[date]* requesting details of the
loss and/or expense which we have incurred to date relating to
our application [to the Architect *(JCT 98)*] of *[date]*, we now
enclose details which we trust will be sufficient to enable you
to [make the ascertainment *(JCT 98)* or make the required
addition to the Contract Sum *(WCD 98)*] required by Clause 26.1
of the Contract. We also enclose the supporting documentation
referred to in the main enclosure.

Yours faithfully,

(5.09.3 cont)

<u>Extended preliminaries</u>

Foreman
Plant, sheds,
  offices
Telephone
Welfare and   4 weeks @ £12420.60   49,682.40
  safety   all as attached
Scaffold hire   build-up
Transport
Insurances
etc.
etc.

Allowance for head office overheads
  and profit 3.1%   1540.15
                                                        £51222.55

[*Note: Other items would be listed and further details
given as appropriate. Supporting documentation must
be included.*]

5.09.4 Letter where the architect seeks to limit the entitlement to extensions granted (JCT 98)

To the Architect   *Date*

Dear Sir,

[*Heading*]

We are in receipt of your letter of [*date*] concerning our
entitlement to recovery of direct loss and/or expense under
Clause 26 of the Contract.

We note that you are seeking to argue that our entitlement must
be limited to the expenses incurred only during the extended
Contract period in respect of which you have granted an
extension of time of 7 weeks. We must point out to you that there
is nothing in the terms of Clause 26 which limits our
entitlement in this way. The Clause quite clearly entitles us
to recover any and all loss and/or expense directly incurred as
a result of the material effect on regular progress of the
Works arising from [*state cause*]. Our entitlement therefore
includes the financial effect of the loss of productivity of
trades engaged upon the Works at the time when the fittings
were to have been installed, as well as the other items listed
and supported by documentary evidence in the further details
which we provided to you.

Further, it was our clearly expressed intention shown in our
master programme to complete the Works four weeks before the
date for completion stated in the Contract. As we made clear to
you when giving our notice of delay under Clause 25, the overall
delaying effect on completion of our work was anticipated to be
11 weeks and this has been borne out by events.

We therefore require the costs of the full period of 11 weeks to
be reimbursed as shown in the details provided.

If you do not issue an extension of time of the length applied
for and include the costs that we have claimed in your next
certificate we shall be forced to issue a Notice of
Adjudication in respect of this matter.

Yours faithfully,

5.09.5 Letter to the architect requesting a breakdown of the extensions of time (JCT 98)

To the Architect *Date*

Dear Sir,

[*Heading*]

Further to your letter of [*date*] granting us an extension of
time of [*specify number*] weeks and our letter of [*date*] making
application for reimbursement of direct loss and/or expense
likely to be incurred as a result of certain of the events to
which the extension also relates, we shall be grateful if, in
accordance with Clause 26.3 of the Contract, you will let us
know what periods of the overall extension relate to the
following relevant events:
Clause 25.4.5.1 Variation and instructions issued by you
                  under Clause 13.2.
Clause 25.4.6   Late issue of instructions.
[*etc. as appropriate.*]

We believe this information to be necessary to enable your
ascertainment of our direct loss and/or expense to be carried
out as required by Clause 26.1.

Yours faithfully,

5.09.6   Letter when ascertainment is delayed

To the Architect or Quantity Surveyor (JCT 98)   Date
(or Employer (WCD 98))

Dear Sir,

[Heading]

Further to your letter of *[date]* in which you indicate that you
do not intend to ascertain the direct loss and/or expense in
respect of which we have made applications to you [the
Architect] of *[date]* and *[date]* until a much later stage, we
cannot accept that this is in accordance with the terms of the
Contract.

The JCT Contract, Clause 26.1 of the 1998 edition, which
applies in this case, requires that following our application
[you shall 'from time to time thereafter' ascertain the amount
of loss and/or expense which has been or is being incurred by us
(JCT 98) or the amount of such loss and/or expense . . . shall be
added to the Contract Sum *(WCD 98)*] subject to certain provisos
which we believe we have met in full. The phrase we have quoted
clearly means that [the ascertainment is to take place *(JCT 98)*
or the payment is to be made *(WCD 98)*] as and when the loss and/
or expense is incurred [and the amounts when ascertained,
whether in whole or in part (see Clause 3), are to be included in
Interim Certificates thereafter as required by Clause
30.2.2.2 *(JCT 98)*].

We shall therefore be glad of your immediate confirmation that
you will [carry out the ascertainment in respect of the amounts
of our loss or expense *(JCT 98)* or make the payments of our loss
and/or expense *(WCD 98)*] so far incurred as notified to you in
our letter of *[date]*.

Yours faithfully,

5.09.7 Letter protesting at the amount ascertained

To the Architect or Quantity Surveyor (JCT 98)   *Date*
(or Employer under WCD 98)

Dear Sir,

[Heading]

With reference to your letter of [*date*] notifying us of the
amount of direct loss and expense [purported to be ascertained
by [the Architect] you for inclusion in the next Interim
Certificate (*JCT 98*) or that you intend to add to the Contract
Sum (*WCD 98*)] we note that this amount is substantially less
than the total of the details of loss and/or expense actually
incurred by us as set out in the annexe to our letter of [*date*]
and supported by the documents supplied therewith. In
particular, the amount in respect of loss of productivity of
labour has been reduced by you by 85 per cent, despite the very
clear evidence that we provided in support of our own figure
and we note that you have not given any justification or any
evidence in support of this reduction or the other reductions
which you have made.

We cannot accept this as being an ascertainment in accordance
with the terms of the Contract and unless the position is
rectified forthwith we shall be compelled to give Notice of
Adjudication [to the Employer (*JCT 98*)].

Yours faithfully,

5.10   Common law claims

Although the architect has no power to settle these claims (*JCT 98*), obviously if a figure can be agreed in respect of them, this is best from every point of view – except that of the legal profession! You should ensure that the architect and/ or the quantity surveyor obtains the employer's express authority to meet and settle common law claims if they are willing to deal with them (**document 5.10.1**).

There is another situation you should guard against, namely that you have waived your common law rights. It is always best to notify a common law claim as soon as possible after it arises, confirming the relevant facts, as memories may well differ some months or years later (**document 5.10.2**).

This letter which relates to a possible JCT 98 scenario could be amended were a similar situation to arise under WCD 98.

A dispute of this nature does not arise *under* the contract and as a result the adjudication provisions of the contract cannot be utilised to resolve the problem. One benefit bestowed by the 1998 editions of the JCT Contracts is the availability of arbitration, or litigation if the arbitration clause in the contract be deleted, at any time. Access to arbitration is no longer prevented prior to practical completion, as was the case for many types of dispute under the earlier edition of the contract.

5.10.1 Letter to the employer where the architect is prepared to settle common law claims (JCT 98)

WITHOUT PREJUDICE

To the Employer   *Date*
(Copy to the Architect)

Dear Sir,

[*Heading*]

Further to the meeting between ourselves and the Architect
held on [*date*] we shall be grateful for your formal
confirmation that the Architect is expressly authorised by
you to negotiate and settle our common law claim arising from
your failure to give possession of the site on the Date for
Possession stated in the Contract, and as notified to you in
our letter of [*date*] which was sent by special delivery.

Yours faithfully,

5.10.2   Notification of a common law claim

SPECIAL DELIVERY

To the Employer   *Date*
(Copy to the Architect)

Dear Sir,

[Heading]

No doubt the Architect will have informed you of the
discussions which have taken place regarding the totally
misleading statements made in the Preliminaries Section of
the Bills of Quantities under the heading <u>Sequence of
Operations</u>, as well as the factual statements made in your
Architect's letters of *[date]* and *[date]* in response to pre-
tender queries which we raised.

Relying on these misrepresentations, we reduced our intended
tender figure by £*[x]* and also reduced the period for
completion from 208 to 195 weeks. We also submitted a programme
which was approved in writing by both your Architect and your
Structural Engineer.

These facts give rise to a claim against you which does not fall
under any provision of the Contract, and we hereby notify you
of our intention to pursue this claim at the earliest
opportunity, if necessary by way of arbitration or
litigation. Our solicitors will be writing to you shortly.

Yours faithfully,

5.11 Reimbursement of loss and expense under IFC 98

The machinery for the recovery of direct loss and/or expense is set out in clause 4.11. This requires you to make written application within a reasonable time of the happening of the event relied on. Essentially, clause 4.11 is a condensed version of JCT 98, clause 26, but there are some significant differences. As under JCT 98, it is not every loss and expense which is recoverable and the general comments in section 5.08 are equally applicable to claims under this clause.

The specified causes are:

- The employer having deferred giving you possession of the site (optional).
- Your not receiving in due time necessary instructions etc. from the architect/contract administrator for which you specifically applied in writing at the correct time. The instructions include those relating to expenditure of provisional sums.
- The opening up for inspection or testing of any work or materials unless the inspection etc. showed that the materials etc. were not in accordance with the contract.
- Execution of work not forming part of the contract by the employer or someone else employed or engaged by him, or failure to execute such work.
- The supply by the employer of materials and goods which the employer has agreed to supply for the works, or the failure so to supply.
- Postponement instructions of the architect/contract administrator under clause 3.15.
- Failure by the employer to give access to the site over any land etc. in his own possession and control.
- Architect's/contract administrator's instructions issued under clause 1.4 (inconsistencies) correcting the quality or quantity of the work or under clause 3.6 (variations) or under clause 3.8 (provisional sums but *not* those for the expenditure of a provisional sum for defined work included in the bills) or under clause 3.3 (named subcontractors) to the extent provided therein.
- Inaccurate approximate quantities. Where an approximate quantity for work is included in the contract documents and it is not a reasonably accurate forecast of the work required.
- Compliance or non compliance by the employer with clause 5.7.1 (relating to planning supervision).
- Suspension by the contractor of the performance of his obligations pursuant to clause 4.4A.

In any such event, you must make a written application to the architect/contract administrator for reimbursement and **document 5.09.1** may be suitably adapted. You are bound to submit in support of your application such information required by the architect/contract administrator or the quantity surveyor as is 'reasonably necessary' to enable the loss and/or expense to be ascertained and **documents 5.09.1** to **5.09.3** are generally applicable and may be adapted as appropriate. The important point is that of your obligation to provide the architect/contract administrator or quantity surveyor with the necessary information; it is not conditional on a request from them.

5.12   Delays and disruptions under the JCT Minor Works Form 1998

The machinery for dealing with delays under the JCT Minor Works Form 1998 is very simply set out in clauses 2.1, 2.2 and 2.3. Clause 2.1 provides for the dates on which the works 'may be commenced' and by which they 'shall be completed' to be set out; clause 2.2 provides for the architect/contractor administrator to grant an extension of time if the works are delayed 'for reasons beyond the control of the Contractor' and clause 2.3 provides for the payment of liquidated damages at a stated rate if the works are not completed by the due date.

The whole concept of 'possession of the site' therefore does not appear in this form. You are entitled to start work on a stated date, but the extension of time clause is so broadly worded that it probably permits the architect/contract administrator to grant an extension if there is any delay in allowing you to start.

Clause 2.2 requires you to notify the architect/contract administrator 'if it becomes apparent that the Works will not be completed by the date for completion inserted in clause 2.1 hereof (or any later date fixed in accordance with the provisions of this clause 2.2) for reasons beyond the control of the Contractor'; **document 5.12.1** is an example of such a notice. The architect/contract administrator is then to 'make, in writing, such extension of time for completion as may be reasonable'. If there is any argument over the architect's/contract administrator's decision a letter similar to **document 5.05.3** may be sent.

There is no provision in the form for recovery of 'direct loss and/or expense' if work is disrupted as there is in JCT 98 and IFC 98. Any claims in this respect must therefore be pursued at common law (see section 5.10) unless the contract has been amended by the inclusion of a 'money claims' clause.

5.12.1 Notification of delay under MW 98, clause 2.2

To the Architect/Contract Administrator   *Date*

Dear Sir,

[Heading]

This is to notify you, as required by Clause 2.2 of the Contract, that it has become apparent that the Works will not be completed by the Date of Completion in Clause 2.1 of the Contract/by the Date for Completion currently fixed following your previous extensions of time granted on *[state dates]* due to *[state reasons, e.g.* exceptionally adverse weather conditions/variation instructions issued under Clause 3.6]*.

[Though the clause does not require you to do so you might go on to suggest the appropriate extension, e.g. in our opinion/ as shown on the attached revised programme we estimate that the delay in completion is likely to be 3 weeks.]

We shall be grateful if you will let us have your written extension of time as soon as possible so that we may proceed to ensure completion by the date then due.

Yours faithfully,

Chapter 6
Completion and Defects Liability

6.01   Practical completion

Nobody knows what 'practical completion' means. JCT 98 and WCD 98 include the words in clause 1.3 'definitions' but all they do is refer to clauses 17.1 and 16.1 respectively in which the words have no definition. JCT 98 and IFC 98 make it a matter for the opinion of the architect and it is within his discretion to decide when 'Practical Completion of the Works is achieved'. Many architects are willing to issue a certificate of practical completion when minor work still remains to be done and may attach a list of such work to the certificate. Others will insist that you complete the whole of the work to be done under the contract, subject only to putting right trivial defects. Strictly speaking, this second view is probably right since the contract really only provides for the putting right of defects appearing after practical completion has been certified. In practice the employer is usually willing to move into the building even though it is not absolutely perfect, but it is clear that the architect cannot issue a practical completion certificate where there are major patent defects in what you have done.

WCD 98 is totally silent as to what practical completion relates to. All that is required is that you provide the health and safety file in accordance with 6A.5.1 if you are the planning supervisor or if 6A.5.2 applies, you provide all the necessary information for the planning supervisor to complete the health and safety file.

Ian Duncan Wallace in the eleventh edition of *Hudson's Building and Engineering Contracts* makes the following contribution to the debate (9.043, page 1130):

> 'It is submitted that this will in the absence of contrary indication, mean that when the work reaches a state of readiness for use or occupation by the owner and free from any known omissions or defects which are not merely trivial.'

It is important to obtain a certificate of practical completion for a number of reasons, one of the most important being to secure the release of half the retention monies. Moreover, if you are in delay in completing, its issue will define the end of the period for which you will have to pay liquidated damages. Again, while the contract does not require you to do so, it is sensible to write to the architect when you believe that the certificate of practical completion should be issued (**document 6.01.1**).

If the architect refuses to issue the certificate, even though you are certain that it should be issued, then your remedy lies initially in adjudication followed by arbitration if the adjudication or threat of it does not sort things out. Before serving notice of adjudication on the employer (see Chapter 9) you might try a letter on the lines of **document 6.01.2**.

6.01.1   Letter giving notice of practical completion

To the Architect   *Date*
(*or Employer under WCD 98*)

Dear Sir,

[*Heading*]

We are pleased to inform you that we have today completed the
Works under the Contract. No doubt you will wish to visit the
site to satisfy yourself that this is so and we look forward to
receiving your Certificate of Practical Completion naming the
day upon which it was achieved.

[*If there are minor defects or minor works to be completed, of
which the Architect is aware, the first sentence should refer
to them, e.g.*
As we know that the Employer is anxious to take possession, we
are pleased to inform you that we have today completed the
Contract Works except for the following minor items and snags
of which you are aware: *list*].

Yours faithfully,

6.01.2 Letter where the architect refuses to certify practical completion

SPECIAL DELIVERY

To the Architect   *Date*

Dear Sir,

[*Heading*]

We have received your letter of [*date*] in which you state that, following your inspection of the Works on [*date*] in your opinion Practical Completion has not been achieved because [*state reasons, if any, given in Architect's letter, e.g.* there are paint splashes on some of the light switches].

Although we are making arrangements to remedy these items, we do not consider that there is a sufficient reason to refuse the issue of the Certificate since these are matters which would customarily be dealt with when rectifying defects.

Unless we receive your Certificate of Practical Completion by return we shall give the Employer formal notice of adjudication. Should this course be necessary, we shall include in the reference those matters regarding extensions of time and reimbursement of direct loss and/or expense which still remain unresolved between us.

We are sending a copy of this letter to the Employer.

Yours faithfully,

This can be amended under WCD 98 to the employer direct

6.02   Sectional completion and partial possession

The JCT 98 contract as printed provides only for a single completion date for the whole of the work. There is no provision for enforceable sectional completion in the articles of agreement. It is possible for the contract bills to require the work to be carried out in a particular sequence and even to state completion dates for individual parts of the work. In such a case, the employer would still only be able to recover any liquidated damages from the single date stated in the appendix, but he might be able to sustain a claim for general damages if you failed to comply with the phased completion dates stated in the bills. If the employer wishes to enforce a series of completion dates with liquidated damages, he must use the Sectional Completion Supplement which formalises the specific provisions that apply to each of a number of defined sections of the overall project. The appendix to the Sectional Completion Supplement provides the opportunity to set out such things as the individual completion date and the liquidated and ascertained damages that will apply to each section.

Where the employer wishes to take possession of part of the work before the whole is completed, he must seek your permission to do so and the consequences are fully set out in clause 18 of JCT 98 (clause 17 of WCD 98 and 2.11 of IFC 98), including adjustment of liquidated damages on the remainder of the work. You do not have an absolute veto because the contract states that your consent 'must not be unreasonably withheld'. It is obviously in your interests, if it does not cause undue inconvenience, to allow the employer to take possession in this way. **Document 6.02.1** is a suitable letter of consent.

You should note that there is no specific provision in the contract for you to recover any additional expense in which you may be involved through the giving of partial possession to the employer. The clause does seem to give you the opportunity to impose conditions upon the giving of consent, which might include the payment of money by the employer, outside the terms of the contract. **Document 6.02.2** is a letter which might be written in these circumstances.

When the employer takes part possession, the architect is required to issue a written statement identifying the relevant part and the date on which the employer has taken possession (you do this yourself under WCD 98). The clause in the contract then operates to relieve you of responsibility for that part of the works, but it does not make a final settlement financially, although it provides for a proportionate reduction in liquidated damages.

There is no provision for partial possession in MW 98. The treatment of sectional completion differs between the forms. JCT 98 has a separate supplement, as does IFC 98. WCD 98 has the provisions printed within the contract document and MW 98 has no such provisions at all.

6.02.1   Letter to the employer consenting to partial possession

To the Employer   *Date*
(Copy to the Architect)

Dear Sir,

[*Heading*]

Thank you for your letter of [*date*] requesting partial possession of [*specify part or parts concerned*].

We hereby give our consent to your so doing subject to the following conditions: [*specify*].

Will you be good enough to arrange for the Architect to issue the necessary written statement identifying the parts you have taken into possession and the date upon which possession is taken.

We would remind you that under Clause 18 of the Contract (*JCT 98*) [(Clause 17 (*WCD 98*) or Clause 2.11 (*IFC 98*)] the insurance for which we are responsible under [Clause 22A] ceases to cover the part in question from the date on which you take possession, and no doubt you will make arrangements accordingly.

Yours faithfully,

6.02.2   Letter to the employer refusing consent

To the Employer   *Date*
(Copy to the Architect)

Dear Sir,

[*Heading*]

We understand that you wish to take possession of Block A and
the surrounding area of the site in accordance with Clause [18
(*JCT 98*) or 17 (*WCD 98*) or 2.11 (*IFC 98*)] of the Contract and
that you are seeking our consent to this.

We very much regret that on this occasion we cannot give our
consent because [*give reason, e.g.* this will create problems
of access which will undoubtedly lead to greater expense and
there is no provision in the Contract under which this could be
recovered.]

[However, if you (or the Architect) are willing to authorise a
variation under the terms of Clause [13.1.2 (*JCT 98*) or 12.1.2
(*WCD 98*) or 3.6.2 (*IFC 98*)] and to reimburse any such expense to
which we become entitled as a result, we shall be pleased to
reconsider the matter.]

Yours faithfully,

6.03   Defects liability

Neither clause 17.2 of JCT 98 nor any similar clause in the other forms is a maintenance clause, although the period to which they relate is often called 'the maintenance period'. You are not required to maintain the works but only to put right any 'defects, shrinkages or other faults' which appear within the defects liability period (which is usually either six or twelve months from the date of practical completion) and which are 'due to materials and workmanship not in accordance with this contract', i.e. as set out in the contract documents or any variation, 'or to frost occurring before practical completion'.

The clauses require the architect to specify the defects in a schedule or notice to be delivered to you not more than 14 days after the end of the defects liability period. You must then make good those defects within a reasonable time at your own cost – provided, of course, that they fall within the definition quoted above.

Defects can appear during the defects liability period, which need urgent attention. Clause 17.3 of JCT 98 empowers the architect to instruct you to put these right at the time.

When you have made good the defects for which you are liable, as specified in the architect's schedule or notice, he must issue a certificate to that effect. It is important that this certificate is issued at the right time since it is one of the events upon which the date for issue of the final certificate depends. It will also be the date following which you become entitled to the release of the balance of retention currently held.

Document 6.03.1 is a letter which you might send if some of the items listed in the schedule of defects are not your liability.

The contract makes no provision for the process known as 'snagging' which usually takes place before the certificate of practical completion is issued. The purpose of snagging is essentially to ensure that the building is reasonably free from defects before the employer takes it over. It may be argued that there is no need for an express contractual provision for this, since your obligation is to carry out the work in accordance with the contract and strictly speaking you should do so, so far as reasonable inspection will reveal, before practical completion is certified (see section 6.01). In your own interests you should not object to making good genuine defects revealed in the process of snagging since, clearly, it will be cheaper for you to do it at that time while you are still on site.

Snagging should only relate to defects which are due to your failure to carry out the work in accordance with the contract.**Document 6.03.1** can be adapted for use where, in your view, the snagging list contains items for which you are not liable.

Although none of the JCT forms expressly require you to notify the architect (employer WCD 98) when you have completed the making good of defects, it is wise to notify him so that the necessary certificate (notice under WCD 98) can be issued and **document 6.03.2** is a suitable form of notification. **Document 6.03.3** is a letter for use where the architect refuses to issue a certificate of making good defects.

MW 98 is far less detailed in its requirements in respect of defects, merely requiring in clause 2.5 that any defects etc. shall be made good and that the architect/contract administrator certifies that this has been done. The examples given can be suitably modified if this contract applies.

Your obligation regarding defects does not end with the issue of the final certificate (final statement under WCD 98). Any latent defects which might subsequently appear, and which are due to your failure to carry out the work in accordance with the contract, will be breaches of contract for which you continue to be liable until the end of the statutory limitation period. You cannot be compelled to return and remedy such defects after the issue of the final certificate, but if they are clearly your responsibility you will be liable for the cost of remedying them. If a claim is made against you, consider carefully whether it is best to offer to return, as the alternative may be far more expensive.

6.03.1   Letter claiming that some items are not defects

To the Architect *Date*

Dear Sir,

[*Heading*]

Thank you for the Schedule of Defects dated [*date*]. We accept
responsibility for the items listed, with the exception of the
following which are not due to materials not being in
accordance with the Contract or to frost occurring before the
date certified for Practical Completion:
[*List excepted items, e.g.*
Defects in plasterwork which are due to normal shrinkage,
etc.]

We are proceeding to rectify all the other items in your
Schedule for which we are responsible under the Contract terms
and we will notify you as soon as they have been made good so
that you may carry out your inspection and issue the
Certificate of Completion of Making Good Defects.

Yours faithfully,

6.03.2   Letter notifying that making good of defects is completed

To the Architect   *Date*
(or Employer under WCD 98)

Dear Sir,

[Heading]

We are pleased to inform you that we have now completed the
making good of all those defects notified to us in your
schedule of defects dated [*insert date*].

We shall be pleased if you can make arrangements for an
inspection as soon as possible so that you may issue the
[Certificate of Completion of Making Good Defects under
Clause 17.4 (*JCT 98*) or certificate that we have discharged our
obligations under Clause 2.10 (*IFC 98*) or 2.5 (*MW 98*) or Notice
of Completion of Making Good Defects under Clause 16.4 (*WCD
98*)] of the Contract.

Yours faithfully,

6.03.3   Letter where the architect refuses to issue a certificate

To the Architect *Date*

Dear Sir,

[*Heading*]

We have received your letter of [*date*] in which you state that you are not yet prepared to issue your Certificate of Completion of Making Good Defects under Clause 17.4 (*JCT 98*) of the Contract because certain alleged defects listed therein have not been made good.

All the defects listed in your letter are among those of which we notified you in our letter of [*date*] and are not due to materials or workmanship not in accordance with the Contract or to frost occurring before the date certified for Practical Completion, and we formally repudiate any liability therefor.

We hope that on reconsideration you will issue your Certificate in accordance with the terms of the Contract, failing which we shall have no alternative but to give a notice of adjudication in accordance with Clause 41.1 (*JCT 98*) of the Contract.

Should the Employer wish us to make good those items listed in your letter, which we hereby dispute, we shall of course be glad to do so [on payment of a sum to be agreed *or* if the Employer will pay us on a daywork basis for the work].

Yours faithfully,

6.04   Final certificate

All the JCT contracts except WCD 98 require the architect to issue the final
certificate within a limited period of time. Under JCT 98, the timetable is
prescribed by clause 30.8. The architect *must* issue the final certificate no later
than two months after the happening of the latest of three events:

- The end of the defects liability period.
- The date on which the architect issued a certificate of completion of making
  good defects.
- The date on which he sent to you a copy of the adjustment of the contract
  sum (clause 30.6.1.2). The architect must send you this adjustment within
  three months after the date of receipt of the necessary documentation from
  you. It is your obligation to see that he has this information within a max-
  imum of six months from practical completion (three months in the case of
  MW 98 unless amended) and he (or the quantity surveyor) is then bound to
  make the necessary computations within three months of receipt of the
  documentation from you. **Document 6.04.1** is the sort of letter you should
  use when sending the information to him. In both cases, your sending the
  necessary information is necessary to enable the architect or quantity surveyor to prepare a final statement of all valuations.

A similar procedure applies under WCD 98 but in that contract you have the
obligation to issue a final statement which the employer has the option to
contest and issue his own final statement.

The final certificate or final statement entitles you to payment of the whole
balance of the contract sum as finally adjusted or, if you have been overpaid,
entitles the employer to recover from you the amount of overpayment. In
either case this sum is expressed as being a *debt* as from the 14th day after issue
of the certificate. It also effectively releases you from any obligation under the
contract, save only the consequences of any breach of contract not discovered
at this date.

Problems do arise over the issue of the final certificate, but case law
establishes that if the architect fails to issue the final certificate as the contracts
require, this is a breach of contract for which the employer will be liable.
Architects' reluctance to issue a final certificate often stems from their misunderstanding of its effect. **Document 6.04.2** is the sort of letter you should
send to the architect if he fails to issue the certificate on the due date. Other
problems can arise over the issue of the final certificate and **document 6.04.3**
deals with a fairly typical situation.

6.04.1 Letter enclosing necessary information for preparation of final adjustment of the contract sum

To the Architect
(or Quantity Surveyor if
so instructed by Architect)   Date

Dear Sir,

[*Heading*]

In accordance with Clause 30.6.1.1 (*JCT 98*) we enclose all the details and information necessary for the adjustment of the Contract Sum and the preparation of the final account. We shall be pleased to provide any further details which you may consider necessary.

No doubt you will now proceed to make the computations so as to enable the Final Certificate to be issued in accordance with the Contract.

Yours faithfully,

6.04.2   Letter where the architect has not issued a final certificate by the due date

---

SPECIAL DELIVERY

To the Architect   *Date*
(copy to the Employer)

Dear Sir,

[Heading]

We have now complied with our obligations under Clause 17 of
the Contract and you issued a Certificate of Completion of
Making Good Defects on *[date]* in which you released the second
half of the Retention.

   The Quantity Surveyor has already ascertained our loss and/
or expense under Clause 26.1 at your instruction. We received
the Quantity Surveyor's statement of adjustments made to the
Contract Sum as is required by clause 30.6.1.2.2 prior to the
issue of your Certificate of Completion of Making Good
Defects. It is now more than 2 months since you issued that
Certificate and the Contract requires you to issue the Final
Certificate. We look forward to receiving that Certificate
within the next 7 days.

Yours faithfully,

6.04.3 Letter where the architect refuses to issue a final certificate

SPECIAL DELIVERY

To the Architect   *Date*
(*Copy to the Employer*)

Dear Sir,

[*Heading*]

Thank you for your letter of [*date*] in which you state that you are unable to issue your Final Certificate because the Quantity Surveyor alleges that he has not yet received all the necessary documentation in accordance with Clause 30.6.1.1 (*JCT 98*) of the Contract.

We must dispute this.

The position is that the defects liability period has elapsed and you have issued the Certificate of Making Good Defects. We submitted the final batch of documents, including the remaining accounts from nominated subcontractors and suppliers to the Quantity Surveyor, with our letter of [*date*], receipt of which was acknowledged by him on [*date*]. Substantially more than the three months allowed to the Quantity Surveyor under the contract to carry out his computations have since elapsed, and in that time we have had no indication from the Quantity Surveyor or from you that any further documents were required. In the circumstances, we must demand immediate issue of the Final Certificate in accordance with Clause 30.8 (*JCT 98*) since the prescribed period of time has now elapsed.

[*Optional:*
We would respectfully point out that your failure to issue the Final Certificate constitutes a breach of contract for which the Employer is liable in damages and we reserve all our rights and remedies in this respect.]

Failing such issue we shall have no alternative but to issue a notice of adjudication/arbitration.

Yours faithfully,

Chapter 7
Determination

7.01 Introduction

Determination of your employment under the contract is a very serious matter, both practically and financially. It is the last resort and a false step is fraught with danger. Wrongful determination of your employment by either you or by the employer could amount to a repudiation of the contract. The party in the wrong might have to pay damages. The detailed rules and procedures laid down by the contract must be followed exactly.

This chapter deals in detail with these rules and procedures.

Alternative specific methods for delivery of the notices required in connection with determination are set out in the contracts. I have only indicated one of these, Special Delivery, at the top of each example in this chapter. The other methods set out in the contracts are, of course, equally valid.

It is, however, worth bearing in mind the provisions introduced into the contract in accordance with the Housing Grants, Construction and Regeneration Act 1996 which allow any dispute to be referred to adjudication at any time. It is suggested, unless the determination is absolutely unavoidable, that the identification of the disputed matters that give rise to the possible determination and their reference to adjudication may avoid the need to determine your employment and all the dangers that this step creates. Even where the possible determination is instigated by the employer, it may be that a referral of the matters giving rise to that threat to adjudication could produce a result that is more in the interests of both parties than the long drawn out problems that will result from a determination.

Suspension of performance is not determination of employment but it is another fallback which can be used but only in cases of non payment (see Chapter 4).

7.02 Determination by the employer

JCT 98 and WCD 98, clause 27, and IFC 98, clause 7.2–7.8, cover the determination of your employment by the employer for reasons which are your fault. Each clause specifies the grounds for determination and the procedure to be followed. The clauses also say what is to happen after determination.

Work related grounds for determination

There are five grounds for determination, namely, if you:

- without reasonable cause *wholly* suspend the carrying out of the works before completion
- fail to proceed regularly and diligently with the works
- refuse or neglect to comply with a written notice from the architect (employer WCD 98) requiring you to remove defective work or improper materials or goods and by your refusal or neglect the works are materially affected
- fail to comply with clause 19.1.1 or 19.1.2 (JCT 98), clause 18.1.1, 18.2.1 or 18.2.3 (WCD 98), or clauses 3.1, 3.2 or 3.3 (IFC 98) (assignment and sub-letting and named subcontractors)
- fail to comply with the requirements of the CDM regulations

In practice, none of these grounds is easy for the employer to establish. You can often successfully challenge a purported notice served by the employer for any of these reasons. Note that the employer's notice must not be served 'unreasonably or vexatiously'.

Wholly suspending the works without reasonable cause

'Wholly' is the operative word. It means 'completely, totally or entirely' and so partial suspension or 'going-slow' is not enough. There must be a complete cessation of work by you or by one of your subcontractors. The stoppage must be 'without reasonable cause'. A 'reasonable cause' might be, for example, if the architect had failed to give instructions, etc. for which you had made specific written application.

Failing to proceed regularly and diligently

Merely failing to comply with your programme is not failure 'to proceed regularly and diligently', although some architects seem to think that it is. This is a very difficult ground for the employer to establish and, depending on the circumstances, you might be able to argue that you had 'reasonable cause' for the alleged failure, for example, if the employer were persistently late in paying certificates, this could well be a 'reasonable cause' for slowing-down so as to reduce your cash-flow problems.

Refusing or neglecting to comply with a written notice from the architect or employer requiring removal of defective work or improper materials or goods and thereby the works are materially affected

Probably, this ground can only be relied on by the employer before practical completion. The works themselves must be materially affected by your refusal or neglect and this is a long-stop provision because the architect can deal with the problem by a correctly-worded instruction under clause 8 (JCT 98 or WCD 98), or clause 3.14 (IFC 98).

Subletting without consent

Whether or not you are in breach of these provisions is a question of fact. You must not, of course, assign or sublet without the appropriate consent under the contract, but that consent is not to be unreasonably withheld.

The employer must follow the determination procedure exactly; if he fails to do so, the determination may be held to be invalid. First, the architect must serve notice on you by actual delivery, special delivery or recorded delivery, specifying the default so as to give you an opportunity of remedying it. If the notice is in general terms, e.g. if it merely alleges that you are not proceeding regularly and diligently without being more explicit, it is probably insufficient. In such a case a letter along the lines of **document 7.02.1** is advisable. It is unwise to take no defensive action even if you think that the architect's allegation is baseless.

In fact, there are three courses open to you:

- You can dispute the notice (**document 7.02.1**) and/or give notice of adjudication (see Chapter 9)
- If you accept that you are in default, you can put matters right and inform the architect that you have done so (**document 7.02.2**)
- You can – unwisely – ignore it and risk the consequences

If you continue the default for 14 days after receiving the notice, then the employer may, within a further ten days, determine your employment under the contract by a properly-worded notice sent by actual, special or recorded delivery. It is only your employment under the contract which is brought to an end; the contract itself remains in existence.

Even if you remedy your default, if you repeat it at any time thereafter the employer becomes entitled to determine your employment without further notice of default from the architect.

The employer's determination of your employment must not be 'unreasonable or vexatious' but you would have to challenge this in arbitration. This is perhaps an area where adjudication may be less than effective in that it could add to the complexity of the situation whichever way the adjudication decision goes if a challenge to that decision is mounted. It is difficult to know what is meant by the phrase 'unreasonably or vexatiously'. *Unreasonably* is a general term which can include anything which can be objectively judged to be unreasonable. *Vexatiously* suggests that there are insufficient grounds and that the notice has obviously been issued simply to cause annoyance. The use of the word 'vexatiously' means that there must be an ulterior motive to oppress or annoy.

Similarly, the employer's determination must not be premature; **document 7.02.3** may be used if appropriate.

Insolvency and corruption

Your insolvency determines your employment automatically because of the provisions of clause 27.3 (JCT 98 or WCD 98) and clause 7.3 (IFC 98). If you become insolvent then, as specified in those clauses, your employment will be automatically determined without need of any notice from either architect or employer. You will know if you are in this situation!

Corrupt practices are a ground on which the employer may determine your employment 'under this or any other contract'. Although the clause is silent on this point this would no doubt be by notice. The wording is extremely wide. (Clause 27.4 (JCT 98 or WCD 98) or clause 7.4 (IFC 98)).

Post-determination procedure

The procedure is laid down in clause 27.6 (JCT 98 or WCD 98), and clause 7.6 (IFC 98) when your employment has been validly determined. In a determination situation you are advised to consult your solicitor.

These clauses lay down the procedure after determination and apply even if you dispute the validity of the determination in adjudication, arbitration or other proceedings, although the clauses envisage that your employment may be reinstated by agreement You are bound to give up possession of the site and, if so instructed in writing by the architect, to remove your temporary buildings, plant, tools, equipment etc. from the works. If you fail to comply with such an instruction within a reasonable time of its being issued, the employer is entitled to remove and sell your property. If he exercises this right he is not liable to you for any loss or damage, but must account to you for the proceeds, after deducting his own costs. You have to wait until completion of the work and the calculation of all final costs before you get any payment. These costs will include any additional expense and direct loss and/or damage which the employer has suffered or incurred as a result of the determination and in practice it is unlikely that you will get anything. Indeed, if there is any adverse balance, the employer may recover this from you.

7.02.1   Letter disputing the sufficiency of a notice of default etc.

SPECIAL DELIVERY

To the Architect   *Date*
(Copy to the Employer)

Dear Sir,

[Heading]

We have received your letter of *[date]* which purports to be a
notice of default under Clause [27.2.1.2 (*JCT 98 or WCD 98*) or
7.2.1(b) (*IFC 98*) *or as appropriate*] of the Contract between us
and *[insert name]*. We deny that we are failing to proceed
regularly and diligently with the Works [*or as appropriate*].
In any case, your notice does not contain sufficient detail for
us or our advisers to understand the nature of your complaint.

Even if, which is denied, there is some lack of progress, this
is the result of your failure to provide [*state relevant item
of information*] for which we made specific written
application to you on [*insert date*].

Will you please provide particulars of the specific default
alleged. In the meantime, we do not consider that your letter
constitutes a valid notice of default under Clause [27.2.1 or
7.2.1 *as appropriate*] and any attempt by the Employer to
determine our employment in reliance on your letter will be
strongly resisted and might well amount to a repudiation of the
Contract.

Yours faithfully,

7.02.2 Letter after putting matters right, if in default

SPECIAL DELIVERY

To the Architect   *Date*
(Copy to the Employer)

Dear Sir,

[*Heading*]

In reply to your letter of [*date*] giving us notice under
[Clause 27.2.1 (*JCT 98 or WCD 98*) or 7.2.1 (*IFC 98*)] of the
Contract that we were in default by [*insert details as set out
in Architect's letter*] we very much regret that this problem
should have arisen.

[*Insert any mitigating factors as appropriate, e.g.* we have
experienced severe mechanical problems with the tower crane
and it is only in the last few days that technicians have been
flown in from Japan to make the necessary repairs.]

We are pleased to confirm that the problem has now been
overcome and that work is now proceeding as required under the
Contract conditions. Indeed we will make every effort to make
up any time which has been lost.

Yours faithfully,

7.02.3   Letter disputing the validity of the employer's notice of determination

---

SPECIAL DELIVERY

To the Employer   *Date*
(Copy to the Architect)

Dear Sir,

[Heading]

We are in receipt of your notice dated [*insert date*] purporting to determine our employment under Clause 27.2.2 (*JCT 98 or WCD 98*) or 7.2.3 (*IFC 98*)] of the Contract.

[*Where notice issued prematurely:*
The Architect's letter giving notice of default was received by us on [*insert date*] as you can confirm with the Post Office. Your notice is dated [*insert date*] which is less than 14 days after the date of receipt by us of the notice of default and is therefore invalid. You should by now have received a copy of our letter [to the Architect (*JCT 98 or IFC 98*)] dated [*insert date*] from which you will see that we have already taken the necessary steps to overcome the problems and so remedy the alleged default.]

[*Where default is denied:*
You will have already received a copy of our letter [to the Architect (*JCT 98 or IFC 98*)] dated [*insert date*] pointing out that the alleged default is denied [*or as appropriate.*] We therefore regard your notice as invalid and are proceeding with the Works.]

Yours faithfully,

7.03 Determination by the contractor

As printed in the contract forms, clause 28 (JCT 98 or WCD 98) and clause 7.9–7.11 (IFC 98) are very favourable to you; they give you powers which you would not otherwise possess. For this reason you will sometimes come across such a clause in an amended form, but if it is to be amended this should have been notified to you in the tender documents. What follows assumes that the clause is unamended.

Before determining your employment under the contract, you should consider very carefully whether it is in your own interest to do so and in any case you should consult your solicitor. A wrongful determination would have serious consequences for you.

There are five main grounds for determination, namely if:

- The employer fails to pay the amount properly due under an Interim or Final Certificate by the final date for payment.
- The employer interferes with or obstructs the issue of any certificate due under the contract (not WCD 98).
- The employer fails to comply with the provisions of the contract regarding assignment.
- The employer fails to comply with the requirements of the CDM Regulations.
- The carrying out of the whole or substantially the whole of the works is suspended for a continuous period of time specified in the appendix to the contract. This is usually one month under JCT 98 and is always one month under IFC 98.
  The suspension must have been due to:
    - failure by the architect/contract administrator/employer to comply with the requirements of the contract relating to information release
    - instructions in relation to inconsistencies, clause 2.3 (JCT 98), clause 2.3.1 (WCD 98), clause 1.4 (IFC 98), variations, clause 13.2 (JCT 98), clause 12.2 (WCD 98), clause 3.6 (IFC 98) or postponement of work, clause 23.2 (JCT 98 or WCD 98), clause 3.25 (IFC 98) unless caused by your negligence or default or that of your subcontractors (other than nominated subcontractors under JCT 98)
    - delays or failure by the employer or others employed or engaged by him on work not forming part of the contract works or in the supply by the employer of goods or materials which he has agreed to supply
    - failure by the employer to give in due time agreed ingress or egress over adjoining property over which he exercises possession and control.

Insolvency of the employer is a further ground on which you can determine your employment in the private edition of JCT 98, and in WCD 98 and IFC 98 all employers are covered.

There must always be two notices from you to the employer (except in case of insolvency when only one is required) – a preliminary notice of default followed by a notice from you to determine your employment.

Employer's failure to pay on time

Adjudication is probably the most appropriate solution in this event as it will probably mean that the outstanding sum is paid quickly, but there may be occasions when a determination is the appropriate remedy.

This is the easiest ground to establish as it is essentially a matter of fact. The words 'properly due' mean that the employer is entitled to deduct any sums which the contract entitles him to deduct, e.g. liquidated damages. The courts have also held that the employer fulfils his contractual obligation by paying the sum actually stated to be due to you by the architect in an interim certificate, even if the certificate contains an error. If you are dissatisfied with the amount stated, you must ask the architect to remedy the error in the next certificate. If he fails to do so, then your proper remedy is to go to adjudication followed by arbitration, or litigation if the arbitration option is deleted in the contract.

This ground, like the other main grounds, is subject to the proviso that you must not exercise your right to determine unreasonably or vexatiously and the courts have said that it would be unreasonable for a contractor to determine his employment where the employer 'is a day out of time in payment, or whose cheque is in the post, or (where) the bank has closed or there has been a delay in clearing the cheque' or something else accidental or purely incidental.

You cannot exercise your right of determination until the final date for payment which itself is 14 days after the date of issue of the certificate. You then give a written notice to the employer (not the architect) (**document 7.03.1**) by actual delivery, special delivery or recorded delivery. This must be done in this way under JCT 98 and WCD 98 and this is also advisable although not compulsory under IFC 98. This notice, if sent by special delivery or recorded delivery, is deemed to have been received 48 hours after the date of posting (excluding Saturday, Sunday and Public Holidays). This delivery is subject to proof to the contrary and, in order to ensure that problems do not arise if delivery is not made by Royal Mail, it is suggested that you make use of the Special Delivery service which means that you can confirm that the notice has been delivered, by telephone any time after one hour after the delivery was due to be made. This can also be done on the Royal Mail website at www.royalmail.co.uk.

This is important because the employer is given a further 14 days after *receipt* of your notice in which to pay the amount due before you become entitled to determine your employment.

Should the employer fail to pay within 14 days of receipt of the first notice, you may determine your employment forthwith or up to 10 days thereafter by a notice (**document 7.03.2**) sent by actual, special or recorded delivery to the employer. The determination takes effect on the receipt of that further notice by the employer.

Employer interfering with or obstructing the issue of any certificate

It has been held that any interference by the employer with the architect's independent duty to issue certificates falls under this head. Such interference would include any attempt to prevent the architect issuing a certificate of any kind, or directing him as to any amount to be included in any money certificate. The problem is particularly acute with certain local authorities whose auditors and councillors have difficulty in appreciating the independent role of the architect under the JCT contracts. You must first serve a notice specifying the default (**document 7.03.3**) and you can only issue a determination notice if the default continues or is repeated.

The employer failing to comply with the provisions of the contract regarding assignment; the employer failing to comply with the requirements of the CDM Regulations

These defaults are of a nature that an appropriate remedy may not be available in adjudication and the only way of dealing with the problem is to determine your employment. **Documents 7.03.1** and **7.03.2** when suitably adapted show the notices that have to be given to determine your employment in these circumstances.

The works or substantially the whole of them being suspended for the period stated in the appendix

The various reasons listed under this head may all be considered as being grounds for which the employer has responsibility or is at fault, e.g. variations or failure to provide necessary instructions. Once the period of suspension stated in the appendix to the contract has elapsed you must give the employer preliminary notice and if the suspension event continues for 14 days you then have to serve the notice of determination not later than 10 days after expiry of the 14 day period. **Document 7.03.4** gives a suitable form of notice. This should be followed up, if the suspension continues for 14 days, by a suitably adapted **document 7.03.2**.

Insolvency of employer

Clause 28.3.1 of JCT 98 (private edition) and WCD 98, and clause 7.10.1 of IFC 98, give you the right to determine your employment on account of the employer's financial failure. You must have sufficient evidence of the employer's insolvency, e.g., a letter from a liquidator or a bankruptcy notice in a newspaper. It is important to note that determination of employment is not automatic, as it would be in the case of your own insolvency (see p. 185). You must serve notice by actual, special or recorded delivery on the employer. This notice determines your employment under the contract forthwith (see **document 7.03.5**). There is no need in this instance for a preliminary notice.

Other rights and remedies

In all cases of determination of employment under the contract – whether by the employer or by you – the printed text states specifically that the contractual right of determination is 'without prejudice to any other rights or remedies' which the party may possess. This preserves any other rights or remedies (for example, an action for repudiatory breach of contract at common law) which the parties have under the contract or at common law and whenever your employment is determined you should seek legal advice.

Reinstatement

If you have properly determined your employment under the contract the employer will be in serious difficulty. It may well be, even at this late stage, that you will be able to reach a satisfactory agreement and continue with the works. You will certainly need professional advice, but if agreement is reached and you agree to go on, the agreement should be confirmed by you as soon as possible and you must be absolutely certain that you and the employer are of like mind (see **document 7.03.6**).

7.03.1   Letter giving notice of default

SPECIAL DELIVERY

To the Employer   Date
(Copy to the Architect)

Dear Sir,

[Heading]

We hereby give you notice under Clause [28.2.1 (*JCT 98 or WCD 98*) or 7.9.1 (*IFC 98*)] of the Contract that you are in default in that you have not paid the amount properly due under Interim Certificate No. 14 issued on [*insert date*] by the final date for payment.

If you continue this default for 14 days from the date of receipt of this notice by you we shall forthwith serve on you a notice determining our employment under the Contract.

Yours faithfully,

7.03.2   Letter determining employment for default

SPECIAL DELIVERY

To the Employer   *Date*
(Copy to the Architect)

Dear Sir,

[Heading]

We refer to our notice dated *[insert date]* and received by you, as notified to us by Royal Mail, on *[insert date]*.

As you have not made payment within 14 days as specified in that notice, we hereby give you notice in accordance with Clause [28.2.3 (*JCT 98 or WCD 98*) or 7.9.3 (*IFC 98*)] of the Contract that we forthwith determine our employment under this Contract.

We are taking immediate steps under Clause [28.4 (*JCT 98 or WCD 98*) or 7.11 (*IFC 98*)] to remove from the site our temporary buildings etc. and all materials and goods properly delivered for the Works and for which you have not yet paid. No doubt you will instruct the Architect in respect of such materials or goods in accordance with Clause [28.4.3.5 (*JCT 98 or WCD 98*) or 7.11.3(e) (*IFC 98*)].

This notice is without prejudice to any other rights or remedies which we may have under the Contract or at common law, and we are placing this matter in the hands of our solicitors.

Yours faithfully,

7.03.3 Letter where interference or obstruction with the certificate is alleged

---

SPECIAL DELIVERY

To the Employer   *Date*
(Copy to the Architect)

Dear Sir,

[Heading]

We have been informed [by the Architect] that the amount stated as due in Certificate No. [*specify number*] issued on [*date*] was reduced by the sum of £[*x*] on the instructions of your Finance Committee.

This constitutes interference with and/or obstruction of the issue of that Certificate. We hereby give the notice required by Clause [28.2.1 (*JCT 98 or WCD 98*) or 7.9.1 (*IFC 98*)] of the Contract that you are in default.

If you continue this default for 14 days from the date of receipt of this notice by you we shall forthwith serve on you a notice determining our employment under the Contract.

Yours faithfully,

7.03.4 Letter of determination following suspension of the works

SPECIAL DELIVERY

To the Employer   *Date*

Dear Sir,

[*Heading*]

As you will be aware, the whole of the Works to be executed under this Contract and so far uncompleted have been suspended for a continuous period of one month since your instruction AI No. 68 of [*date*] postponing the work because [*give reason*].

This letter gives you formal notice pursuant to Clause 28.2.3 [*or as appropriate*] of the Contract that if this suspension continues for 14 days from the date of receipt of this notice by you we shall forthwith serve on you a notice determining our employment under the Contract.

Yours faithfully,

7.03.5   Determination on the grounds of the employer's insolvency

---

SPECIAL DELIVERY

To the Employer   Date
(Copy to the Architect)

Dear Sir,

[Heading]

We have received a letter from Messrs [XYZ] of [address]
stating that [give details of insolvency situation].

In accordance with Clause [28.3.3 (JCT 98 or WCD 98) or 7.10.3
(IFC 98)] we hereby give you notice that we forthwith determine
our employment under the Contract.

[Paragraphs from document 7.03.2 may also be included.]

Yours faithfully,

7.03.6   Letter confirming reinstatement after determination

SPECIAL DELIVERY

To the Employer   *Date*
(Copy to the Architect)

Dear Sir,

[Heading]

This is to confirm the agreement reached at our meeting on
[insert date] held at the Architect's office when, in
consideration of our withdrawing our notice of determination
and resuming work, it was agreed as follows:

[Set out exactly what was agreed in numbered paragraphs, e.g.
1.   No later than *[date]* you will pay us the sum of £*[x]* as
damages in respect of the cause(s) of determination.
2.   The date for completion under the Contract will now be
[date] and you waive all or any right to enforce liquidated
damages against us in respect of any earlier date.
3.   All future sums certified as due to us under Clause *[30 (JCT
98 or WCD 98) or 4 (IFC 98)]* shall bear interest at 5% above Bank
base rate from the date of issue of the Certificate unless that
sum is received by us and cleared by your Bank within 14 days
thereof.
4.   Your solicitors are to prepare an appropriate Addendum
agreement, to be annexed to the Contract.*]*

*[We undertake to withdraw our notice of determination and to
resume work on site within [x] days of receipt by us of the sum
stated in paragraph 1.]*

Yours faithfully,

7.04 Determination by the employer or contractor: no-fault determination

Clause 28A of JCT 98 and WCD 98 and clauses 7.13 and 7.14 of IFC 98 deal with the determination of your employment by you or by the employer where the carrying out of the whole, or substantially the whole of the uncompleted works (other than a suspension caused by your failure to rectify defects under clause 17 (JCT 98), 16 (WCD 98) or 2.10 (IFC 98)) is suspended for a continuous period because of specified events which are beyond the control of either you or the employer. Under JCT 98 and WCD 98 the period of suspension must be stated in the appendix while IFC 98 prescribes periods of three months and one month depending on the reason for the suspension.

The specified events are:

- Force majeure.
- Loss or damage to the works caused by the specified perils such as damage by fire. (You cannot rely on this ground where the loss or damage was caused by your negligence or default or that of one of your subcontractors, but this would not prevent the employer determining your employment on this ground.)
- Civil commotion.

(The above events require a three month suspension under IFC 98 before determination can be effected.)

- Delay in receipt of any permission or approval for the purposes of Development Control Requirements (WCD 98 only).
- Architect's/contract administrator's/employer's instructions in respect of inconsistencies, variations or postponement resulting from the negligence or default of any local authority or statutory undertaker executing work solely in pursuance of its statutory obligations.
- Hostilities involving the UK.
- Terrorist activity.

(The above events require a one month suspension under IFC 98 before determination can be effected.)

If you wish to determine your employment under these provisions only one notice from you to the employer is required, there being no warning or preliminary notice. The determination notice takes effect seven days after the date of receipt of the notice. However, you must not give the notice unreasonably or vexatiously and it must, under JCT 98 and WCD 98, and should under IFC 98, be sent by actual, special or recorded delivery: see **document 7.04.1**.

Where your employment is brought to an end under clause 28A of JCT 98 or WCD 98 or clause 7.13 or 7.14 of IFC 98, the contracts equitably provide that your rights and duties (and the employer's) after determination are the same as when you have determined your employment because of the employer's default except that the employer does not pay for 'any direct loss and/or damage' caused to you by the determination, i.e. you do not get loss of profit on the uncompleted contract.

7.04.1   Letter determining employment forthwith

SPECIAL DELIVERY

To the Employer   *Date*
(*Copy to the Architect*)

Dear Sir,

[*Heading*]

As you are aware, the whole of the Works to be executed under
this Contract and so far uncompleted have been suspended for a
continuous period of [*under JCT 98 and WCD 98 insert period
specified in the Appendix entry; under IFC 98 insert 'three
months' or 'one month'*] because [*insert appropriate dates and
details, e.g.* of the serious fire which occurred on [*date*].

Accordingly, under Clause [28A.1 (*JCT 98 or WCD 98*) or 7.13.1
(*IFC 98*)] of the Contract we hereby forthwith determine our
employment.

We are taking immediate steps under Clause [28A.3 (*JCT 98*) or
28A.4 (*WCD 98*) or 7.16 (*IFC 98*)] to remove all our temporary
buildings, plant etc. and all materials and goods properly
delivered for the Works and for which you have not yet paid from
the site and no doubt the Architect will be dealing speedily
with the financial consequences of the determination of our
employment.

This notice is without prejudice to any other rights or
remedies which we may have.

Yours faithfully,

7.05 Determination under the Minor Works Form

This is covered by clause 7 and the grounds and procedure are different to those under the other JCT contracts considered earlier.

Determination by the employer

There are only two grounds on which the employer can determine your employment under the contract, but you should note that the employer's right is 'without prejudice to any other right or remedy which the employer may possess', i.e. for repudiatory breach at common law.

The two grounds specified are if you:

- Fail to proceed diligently with the works, wholly suspend the carrying out of the works before completion or fail to comply with the CDM Regulations – in all cases 'without reasonable cause'.
- Become insolvent in the way specified in clause 7.2.2.

In the first of these grounds you must first receive a preliminary notice and if the default continues for seven days your employment may be determined by further notice.

In the second of these events, there is no preliminary notice requirement, but merely service on you of a notice of determination which takes effect forthwith.

These notices must be served by actual, special or recorded delivery. If this happens you should consult a solicitor at once, but assuming that the determination is valid, you must leave the site immediately and will receive no further payment until after completion of the works.

Determination by the contractor

There are five grounds on which you are entitled to determine your employment under clause 7.3. In all cases the right must not be exercised 'unreasonably or vexatiously'.

The first four grounds require you to serve a preliminary notice of default. They are, if the employer:

- Fails to make progress payments by the final date for payment.
- Obstructs the carrying out of the works or fails to make the premises available on the commencement date, either himself or through the agency of anyone for whom he is responsible, e.g. the architect.
- Suspends the carrying out of the works for a continuous period of at least one month.
- Fails to comply with the requirements of the CDM Regulations.

You must first serve a preliminary notice of default by actual, special or recorded delivery. The notice must specify the default (see **document 7.05.1**). If the default is continued for seven days after receipt of your notice, you may then forthwith determine your employment by notice served on the employer, again by actual, special or recorded delivery (**document 7.05.2**). Before doing so, you would do well to consult your solicitor.

The fifth ground is if the employer becomes *insolvent* as specified in the clause. This merely requires notice of determination.

In all cases, your right to determine your employment is 'without prejudice to any other rights or remedies' which you have, leaving you free to proceed at common law. If you validly determine your employment, the clause lays down your entitlement. You are entitled to be paid:

- A fair and reasonable sum for the value of work begun and executed.
- A like sum in respect of materials on site.
- The same in respect of cost of removal of all temporary buildings, plants, tools and equipment.
- Any direct loss and/or damage caused to you by the determination.

7.05.1 Letter giving notice of default under MW 98

SPECIAL DELIVERY

To the Employer   *Date*
(Copy to the Architect)

Dear Sir,

[Heading]

We hereby give you notice under Clause 7.3 of the Contract that you are in default in the following respect *[specify default giving necessary details and following contract wording]*.

If you continue the default for 7 days after receipt of this notice we may forthwith determine our employment under the Contract by serving a further notice on you.

Yours faithfully,

7.05.2   Letter giving notice of determination under MW 98

SPECIAL DELIVERY

To the Employer   *Date*
(Copy to the Architect)

Dear Sir,

[Heading]

We refer to the notice dated *[date]* sent to you by special delivery/recorded delivery.

As you have continued the default specified therein for 7 days after its receipt, in accordance with Clause 7.3 of the Contract we hereby forthwith determine our employment thereunder, without prejudice to any other rights or remedies which we may possess.

Yours faithfully,

Chapter 8
Subcontractors and Suppliers

8.01 Introduction

This chapter deals with subcontractors, nominated, named and domestic, and suppliers, nominated or otherwise. It also extends to cover sub-subcontractors, but in general you will not be concerned with these except as regards giving your consent to any subletting by your subcontractors or, of course, if their work etc. proves to be defective.

JCT 98 and WCD 98 each have a separate clause entitled 'Assignment and sub-contracts'. In IFC 98 and MW 98 these matters are dealt with as part of a clause entitled 'Control of the Works'. JCT 98 has very complex provisions relating to nominated subcontractors. IFC 98 provides for 'named' sub-contractors who are selected by the employer to carry out part of the work. WCD 98 has similar provisions but they are within the 'Supplementary Provisions' to this contract which are an optional inclusion. MW 98 has no specific provisions regarding subcontracts and all such subcontracts will be on a domestic basis.

One of the major practical problems is the way in which the main and subcontracts fit together. This affects nominated, named and domestic sub-contractors and the special forms of subcontract have been drafted for the purposes of conformity.

Under JCT 98 there is a series of specific documents relating to nominated subcontractors. NSC/A, the Standard Form of Articles of Nominated Sub-Contract Agreement, and NSC/C, the Standard Conditions of Nominated Sub-Contract, are the ones that apply to your agreement with the nominated subcontractor. Other forms relate to the tendering process (NSC/T Parts 1–3) and to an agreement between the employer and the nominated subcontractor (NSC/W), and there is also a Standard Form of Nomination Instruction (NSC/N) which you will receive from the architect when a nomination is made.

The Construction Confederation produces the standard subcontracts for use with the JCT forms. At the time of writing, DOM/1, which was drafted for JCT 80, is still in its 1980 edition but has been updated to fit in with Amendment 18 of JCT 80. JCT 98 was produced to tidy things up and incorporate Amendment

18 within the text of the contract form. There were some minor drafting amendments when JCT 98 was finalised but nothing of particular note by way of amendment. DOM/1 1980 edition as published in 1998 can be used with JCT 98 although care should be taken to ensure that your subcontracts require subcontractors to comply fully with all applicable main contract conditions.

IFC has an accompanying subcontract for named subcontractors (NAM/SC). Under this contract, form NAM/T is used for tendering purposes. For domestic subcontracts under IFC 98 the Construction Confederation publishes form IN/SC.

WCD 98 does not include any specific requirements as to the form that the subcontract with a named subcontractor should take. The Construction Confederation produces DOM/2 for use in domestic subcontracts under this form.

8.02   Assignment

All the contracts prevent either you or the employer assigning your rights under the contract to a third party, without the consent of the other. The general rule of law is that neither party is entitled to assign his obligations or liabilities under the contract without the other's consent. Clauses 19.1.1 (JCT 98), 18.1.1 (WCD 98), 3.1 (IFC 98) and 3.1 (MW 98) are the applicable provisions.

The only 'right' which you may wish to assign to a third party is the right to receive payment so as to obtain finance from a factoring house. This course is not recommended and in any case you need the employer's consent, which can be refused without any reason being given. Understandably, most employers are loath to grant this consent, but if you need to seek finance in this way, a letter along the lines of **document 8.02.1** is necessary.

8.02.1 Letter to the employer seeking permission to assign the right to payment

---

SPECIAL DELIVERY

To the Employer   *Date*
(Copy to the Architect)

Dear Sir,

[Heading]

Due to a financial restructuring of this Company it has become necessary for us, on all our current contracts, to assign our right to payment thereunder to the factoring house of [*XXXXX*] Ltd.

As required by Clause [19.1.1 (*JCT 98*) etc.] of our Contract with you, we formally request your permission so to assign our rights to payment under the Contract to that firm. We shall be glad to receive from you formal permission as soon as convenient.

Yours faithfully,

8.03   Sub-letting: domestic subcontractors

On most contracts you will wish to sublet some of the work to domestic subcontractors and, indeed, you will have obtained tenders from proposed subcontractors as part of your pre-contract task. You cannot sublet in this way without the written consent of the architect. **Document 8.03.1** is an example of the sort of letter you might send seeking the architect's (or employer's (WCD 98)) consent.

You may also encounter the situation under JCT 98 where the employer has required a list or lists of prospective subcontractors for a certain part or parts of the works to be included in the contract bills. These subcontractors will ultimately be engaged as domestic subcontractors. You are entitled to add further names to the list either of your own volition or to fill a vacancy, and **document 8.03.2** is an example of an appropriate request in these circumstances. There may well be more than one of these lists when the employer wishes to have some say on which firm shall be used for various aspects of the works.

The architect cannot withhold his consent *unreasonably*, and while any refusal of consent is open to adjudication or arbitration, this is not really a practical remedy. If, in your view, the architect has been unreasonable then you can try a letter along the lines of **document 8.03.3**.

Assuming that consent is given, then you must see that the subcontract documentation is completed. In fact, there is an earlier problem with which you may have to deal in connection with both proposed domestic subcontractors and proposed suppliers.

Where your main contract is on JCT 98 terms, you should have required tenders on the basis that the subcontractor will contract with you on the appropriate domestic subcontract, i.e. DOM/1 for JCT 98 and DOM/2 for WCD 98, or, in the case of a supplier, on terms which do not conflict with your liabilities to the employer. Under IFC 98, you should have required tenders on the basis that the subcontract will be on IN/SC terms. However, it is very common for tenders to be submitted with the subcontractor's own conditions and these, in all probability, will attempt to limit that firm's liability in various ways.

If the tender is improperly submitted to you in this way you can either reject it and go elsewhere, or, if this is impossible, try to get the tender re-submitted so as to achieve the desired result (**document 8.03.4**).

This problem is called the 'battle of the forms' and, as regards suppliers particularly, you may find that the small print is headed by or includes a clause saying 'All orders are accepted only upon and subject to the terms set out in this quotation and the following conditions. These terms and conditions shall prevail over any terms and conditions in the buyer's order'. You must ensure that your terms and conditions prevail and as a general rule if you get the 'last shot' in (**document 8.03.5**) you will achieve that result. If there is an acknowledgement slip attached to the quotation it is best not to return it or, if you do, ensure that it is amended appropriately. In the case of supply contracts, you should watch out for clauses which attempt to exclude the seller's liability for defective goods or workmanship, or to retain title in the goods and materials sold.

8.03.1   Letter seeking consent to sub-letting

To the Architect (JCT 98, WCD 98 or MW 98)   Date
(or Employer under WCD 98)

Dear Sir,

[Heading]

In accordance with Clause [19.2 *(JCT 98) etc.*] of the Contract
we shall be glad to receive your written consent to sublet the
following parts of the Works (in accordance with our
notification at the time of tender).

Work	*Proposed subcontractor**
Brickwork	Stoneleight Bricklayers Ltd
Joinery	Perfectfit Joinery Ltd
[etc.]	

Yours faithfully,

**Note: The contract does not require you to notify the
architect of the identity of your proposed subcontractors,
though it would seem only reasonable for you to do so. This
column may therefore be omitted.*

8.03.2   Letter seeking to add an additional name to a list of prospective subcontractors (JCT 98)

To the Architect   *Date*

Dear Sir,

[Heading]

In accordance with Clause 19.3.2.1 of the Contract we shall be glad to receive your written consent to add the following name to the list of firms included within the Bills of Quantities to carry out the brickwork:
Stoneleigh Bricklayers Ltd.

[Alternative to fill a vacancy:
We have to inform you that one of the firms listed in the Bills of Quantities as a prospective subcontractor for the brickwork, Waterside Brickwork, has declined to submit a tender. In accordance with the requirements of Clause 19.3.2.2 we propose the addition of Stoneleigh Bricklayers Ltd to the list.]

We look forward to receiving your written consent to this proposal at your earliest convenience.

Yours faithfully,

8.03.3 Letter protesting at the architect's refusal of consent to sub-letting

SPECIAL DELIVERY

To the Architect   *Date*

Dear Sir,

[*Heading*]

Following receipt of your letter of [*date*] we must protest at your refusal to give consent to the subletting of the brickwork.

We made it clear to you at the time of acceptance of the invitation to tender for this project, and again at the time of tender, that we did not currently have the capacity to carry out brickwork of the kind required within our own organisation and it would therefore be necessary for us to sublet this element of the Works. We therefore consider your refusal to give consent unreasonable and must ask you to reconsider as continued refusal will render further work on this Contract impossible for us. In view of the urgency of the matter please let us have your written decision by return of post.

Yours faithfully,

8.03.4   Letter to the subcontractor regarding unacceptable terms of tender

---

To the Subcontractor   *Date*

Dear Sirs,

[*Heading*]

With reference to your quotation dated [*date*] we must point out
that our invitation to you to quote for the painting work on
this Contract clearly stated that your subcontract with us, in
the event of acceptance of your quotation, would be on the
terms of the Domestic subcontract [DOM/1 (*JCT 98*) *etc.*].

The terms attached to your quotation are inconsistent with the
terms of that subcontract and are therefore unacceptable to
us. Please re-submit your quotation (which is otherwise
acceptable to us) making it clear that the terms of DOM/1 will
apply where they conflict with the terms printed on the back of
your quotation form.

Yours faithfully,

8.03.5   Last shot in the battle of the forms

To the Builders' Merchant   *Date*

Dear Sirs,

[*Heading*]

With reference to your 'Acceptance of Order' in respect of the
supply of roof tiles for this Contract dated [*date*] we repeat
that your quotation dated [*date*] is accepted subject to the
terms and conditions specified in our Order dated [*date*] and
only to those terms and conditions.

As previously agreed the first delivery of 10,000 tiles must be
made to the site on [*date*].

Yours faithfully,

8.04 Letters of intent to subcontractors or suppliers

During negotiations for the letting of a main contract you may need to have certain preliminary work done by proposed subcontractors or suppliers, though this is not usual. If this is necessary, the 'letter of intent' comes into its own, but you must be very careful.

Depending on its wording, a letter of intent may in fact create legal liability so that if, for some reason, the main contract does not come into being, you may nevertheless be liable to the proposed subcontractor or supplier.

From the point of view of the former, it is fair that he should be paid for any preliminary work in any case, and it is best to negotiate to pay him a specified amount if the main contract does not go ahead. In the case of suppliers, of course, all you will have done is to make an inquiry as to availability and price, unless the materials involved are subject to long delivery in which case the same concerns apply. **Documents 8.04.1** and **8.04.2** are examples of letters of intent; the first, if acted on, will create no liability. The second letter imposes a limited liability.

Where a letter of intent issued to you by an employer imposes a liability on you, it would be foolhardy not to have a similar letter in place covering your position with the employer should the project not proceed and you are left with a liability that you would otherwise have to honour without equivalent recompense.

Document 8.04.3 is a suitable first letter in these circumstances and **document 8.04.4** is a response where the employer refuses to play ball.

8.04.1   Letter of intent to the subcontractor (no liability)

To the Subcontractor   *Date*

Dear Sirs,

[*Heading*]

With reference to your quotation dated [*date*] we intend to
enter into a subcontract with you for the supply, delivery and
erection of structural steelwork for this contract in
accordance therewith.

It will be necessary for the erection of steelwork to commence
on site no later than [*date*]. We shall have no objection should
you wish to commence fabrication in anticipation of our
entering into a subcontract with you for this work.*

Yours faithfully,

**Ideally, the letter should make clear that the sender accepts
no financial or other liability should the subcontract not
eventuate. Such a bald statement might put the recipient off!*

8.04.2   Letter of intent to the subcontractor (with limited liability)

SPECIAL DELIVERY

To the Subcontractor   *Date*

Dear Sirs,

[*Heading*]

With reference to your quotation dated [*date*], we intend to
enter into a subcontract with you for the supply, delivery and
erection of structural steelwork for this contract in
accordance therewith.

Since it will be necessary for the erection of steelwork to
commence on site no later than [*date*] please commence
fabrication in anticipation of our eventual execution of a
formal subcontract. Should it prove impossible, for any
reason, for us to enter into a subcontract with you, in that
event we agree to pay any reasonable net costs incurred by you
in reliance on this letter less the sale value of any
materials, whether fabricated or otherwise, to which such
payment relates.

Yours faithfully,

8.04.3 Notification to employer of need to commit to a subcontractor

---

To the Employer   *Date*
(Copy to the Architect)

Dear Sir,

[*Heading*]

In order to fulfil the requirements of the proposed programme
for the above we need to place an order for the steelwork no
later than the end of next week.

As you are aware we have no commitment from you in respect of
this project as yet. We require that you will undertake to
defray all costs that we may incur in respect of any order that
we place for the steelwork.

Please would you issue this confirmation to us no later than
[*date*] in order that we may place the order and ensure that the
currently proposed commencement date is met.

Yours faithfully,

8.04.4 Letter where no or inadequate response to request for a commitment

---

To the Employer   *Date*
(Copy to Architect)

Dear Sir,

[*Heading*]

We have heard nothing from you in respect of our letter of [*date*]. [We have read your letter in reply to ours of [*date*] and we are not satisfied with the terms of the undertaking given.]

Unless we receive your undertaking to reimburse all costs that we may incur as a result of placing an order for the steelwork before we have reached a final agreement on the terms of the contract with yourselves, we shall be unable to comply with the programme as currently proposed and we shall have to delay commencement of our works on site.

Yours faithfully,

8.05 Signing the subcontract

Where part of the works is sublet to a domestic subcontractor, it is the contractor's responsibility to prepare the subcontract documents. The Construction Confederation Domestic Sub-Contract DOM/1 1980 was specially designed for use where the main contract is in JCT 80 form. It has been updated at the time of writing to line up with Amendment 18 of JCT 80. It is therefore not absolutely on all fours with JCT 98. Until the Construction Confederation produces an edition that ties up with JCT 98 care should be taken to ensure that there are no differences which will cause problems later. The Articles of Agreement and the sub-contract conditions are issued as separate documents and only the Articles of Agreement will be used, unless the main and subcontractor have otherwise agreed.

Under IFC 98, the appropriate form of domestic subcontract is IN/SC. At the time of writing IN/SC is in its 1985 edition amended to include Amendment 12 of 1998 and the same warning applies as for DOM/1.

The appendix to these contracts must be carefully completed and the guidance notes printed on the forms should be studied. Do not make any other deletions or amendments to the printed form and take care to ensure that, when completed, it ties up with the appendix in the main contract. The information in the appendix should have been notified to the proposed subcontractor at the time of tendering and his tender will have been submitted to you on that basis. If there are differences you can be almost certain that problems will arise in getting the subcontract signed.

All deletions and amendments should be initialled by both parties at the time the document is signed. If the main contract has been executed as a deed, then so should the subcontract. The only practical difference between a contract merely signed by the parties and one under seal is that the limitation period is 12 years in the case of deeds and 6 years in other cases. The appropriate attestation clause should be used.

These comments apply generally to the use of any standard form of subcontract, including DOM/2 and IN/SC.

8.06 Nominated suppliers

You are entitled to an extension of time under JCT 98, clause 25.4.7 for 'delay on the part of nominated ... suppliers' which you have taken all practicable steps to avoid or reduce. The JCT published two forms for use with JCT 80 where the architect receives tenders from nominated suppliers.

These forms were the Form of Tender for Nominated Suppliers (TNS/1) and the Warranty by a Nominated Supplier (TNS/2). At the time of writing these forms have not been updated to fit in with JCT 98. There is nothing to prevent an architect from using the JCT 80 version to obtain tenders from proposed nominated suppliers even when he uses JCT 98. You should check very carefully if this is done. Clause 36.4 of JCT 98 puts specific requirements on the architect when nominating a nominated supplier and you would be quite entitled to decline the nomination if the tender of the proposed supplier does not comply with clause 36.4

Form TNS/1 is not mandatory in any event and the checking of terms under which a proposed nominated supplier has tendered is even more important in this event.

TNS/1 includes at schedule 3 a warranty (TNS/2) which establishes a contractual relationship between the supplier and the employer. This in no way affects your rights against the supplier if things go wrong, because the provisions of the Sale of Goods Act 1979 apply to the contract of sale between you and him. However, as in the case of subcontractors, you should take particular care to ensure that the supplier's own conditions do not conflict with your obligations under the main contract, and that he does not attempt to limit his liability to you in any way.

In theory you are protected by the provisions of clause 36.4 of JCT 98 which limits the architect's right to nominate suppliers to firms who will accept, as part of the contract of sale, the terms set out in that subclause and it is in your interest to ensure that they do so. But the subclause starts with the statement 'save where the Architect and the Contractor shall otherwise agree'. In some circumstances it may be to your advantage to agree, possibly subject to some conditions; **document 8.06.1** is an example of a letter which might be used.

A possible pitfall for you lies in the practice of some architects who, after you have contracted with the nominated supplier, attempt to vary the goods or materials to be supplied by contacting the supplier directly. You should insist that these matters are dealt with by you (see **document 8.06.2**).

Problems are avoided if the TNS/1 tender form is used. Watch out for any attempt by the supplier to alter the form or introduce his own terms of sale, e.g. by attaching a quotation on a printed form. Write to the offender at once, sending a copy of the letter to the architect (**document 8.06.3**).

You can protect your position where the nominated supplier's quotation contains 'restrictions, limitations or exclusions' which fall outside the terms of clause 36.4. If you note any such restrictions etc., for example, a supplier's term of sale which restricts his liability, you should write to the architect at once (**document 8.06.1**).

The architect can then decide whether by clause 36.5.1 to change his nomination or confirm it. If he decides to confirm it then he must give his specific written approval of the restriction etc. to you.

Note that, unlike the case of nominated subcontractors, you have no right to make reasonable objection to a proposed nominated supplier if, for example, you have had bad past experiences with him. You can only object and refuse the nomination on the ground that the supplier will not sell on the terms set out in clause 36.4. Provided that those terms are not compromised there is of course no restriction on the incorporation of other terms in the contract with the nominated supplier.

8.06.1 Letter regarding non-standard terms of quotation from a nominated supplier (JCT 98)

To the Architect   *Date*

Dear Sir,

With reference to your AI No. 597 instructing us to place an order with Bildrite (Builders' Merchants) Ltd for the supply of roof tiles against the PC sum in item A on page 134 of the Contract Bills, we must point out that the terms of their quotation conflict with the terms set out in Clause 36.4 of the Contract in that, contrary to Clause 36.4.2, their liability in respect of defective materials is limited to their replacement only.

We shall be prepared to accept this nomination only on the following terms:

1.   That we receive your specific approval in writing as required by Clause 36.5.1 to this restriction of their liability so that our liability to the Employer under the Contract in respect of the goods to be supplied by them is restricted, limited or excluded to the same extent.

2.   That we receive a written indemnity from the Employer agreeing to reimburse to us any expenses reasonably incurred by us as a direct consequence of any defective materials supplied subject to the same provisos as are set out in Clause 36.4.2 of the Contract.

On receipt of your specific written approval and the Employer's indemnity as set out above we shall be prepared to accept the nomination and enter into a Contract of sale with the firm concerned.

Yours faithfully,

8.06.2 Letter to the architect regarding his direct contact with a nominated supplier (JCT 98)

To the Architect *Date*

Dear Sir,

[*Heading*]

It is only on receiving delivery of the first load of roof tiles from Bildrite (Builders' Merchants) Ltd that we have learned that, by direct contact with them, you have changed the type of tile from [*specify*] to [*specify*].

Such direct contact with a supplier, without reference to us, makes administration of this contract extremely difficult. We must also point out that the firm's contract of sale is with us, and you have no legal right to change its terms. We must therefore insist that, in future, any such changes in your requirements are addressed in the first instance to us in the form of proper instructions issued under the appropriate Clause of the Contract, which in this case is Clause 13.2. Will you please immediately issue such an instruction in respect of the alteration in the type of roof tile.

Yours faithfully,

8.06.3   Letter to a nominated supplier regarding his attempt to change the terms of form TNS/1

To Bildrite (Builders' Merchants) Ltd   *Date*
(Copy to the Architect)

Dear Sirs,

[*Heading*]

We have been instructed by the Architect to enter into a contract of sale with you as nominated suppliers for the supply of roof tiles for this project on the basis of Form TNS/1 completed by you and of your quotation dated [*date*] attached thereto.

We note that paragraph 9 of the Conditions printed on the reverse of your quotation conflicts with Clause 36.4.2 of the Conditions of the main Contract as printed on page 4 of Form TNS/1 in that it seeks to restrict your liability for any defects in the materials supplied to their replacement only. We must draw your attention to Clause 1.2 on page 1 of Form TNS/1 which states that the contract of sale is to be on the terms set out in Clause 36.3 to .5 of the Conditions of the main Contract and your conditions of sale 'in so far as they do not conflict with the terms of' that Clause. Before concluding our contract of sale with you we shall be glad to receive your written assurance that your liability for defective materials will be in accordance with Clause 36.4.2 of the main Contract and will not be restricted in the manner set out in paragraph 9 of the Conditions on the reverse of your quotation.

Yours faithfully,

8.07 Nominated subcontractors under JCT 98

Under JCT 98 there are extremely complex procedures for the nomination of subcontractors in which you are intimately involved. The documentation to be used depends on which method of nomination is chosen by the architect.

The procedure is set out in clauses 35.6 to 35.9 and involves a good deal of work on your part. You will receive from the architect an instruction on Standard Form of Nomination Instruction (NSC/N).

This will be accompanied by Parts 1 and 2 of NSC/T, the Standard Form of Nominated Sub-Contract Tender. Part 1 will have been completed by the architect. Part 2 will have been completed and signed by the subcontractor and signed by or on behalf of the employer as 'approved'. There will be certain documents, the 'numbered tender documents', listed in and enclosed with NSC/T Part 1 and there may also be additional documents which have been approved by the architect.

Clause 35.6 also requires the following to be attached:

- A copy of the completed NSC/W, which is a direct agreement between the employer and the nominated subcontractor.
- Confirmation of any alterations to the information given in NSC/T Part 1 relating to:
  - Item 7 Obligations or restrictions imposed by the employer
  - Item 8 Order of works: employer's requirements
  - Item 9 Type and location of access.
- A copy of your health and safety plan.

A copy of the instruction is sent to the nominated subcontractor by the architect together with a copy of the completed appendix for the main contract.

You must then carefully check over NSC/T Parts 1 and 2 as completed and you must complete NSC/T Part 3 by negotiation with the subcontractor. Assuming you are able to agree everything, both you and the subcontractor must sign and date form NSC/A, the Articles of Nominated Sub-Contract Agreement, making sure that any deletions or amendments are initially jointly. You then send a copy of the completed and signed NSC/A and the agreed and signed NSC/T Part 3 to the architect. You are not required to send him anything else. The nominated subcontractor should also be sent a completed NSC/A and NSC/T Part 3.

This sounds simple enough, despite the complex nature of the form which even experienced architects and other professionals find difficult to understand. However, you have only *ten working days* from the date on which you receive the preliminary notice of nomination in which to get to grips with the problem. If you and the subcontractor cannot reach agreement within that period, you must give notice in writing to the architect either under clause 35.8.1 telling him the date by which you expect to complete the negotiations with the subcontractor and execute Agreement NSC/A (**document 8.07.1**), or identifying in accordance with Clause 35.8.2 the matters that have caused the non compliance with the requirement to reach agreement and execute NSC/A within the ten day period set out in clause 35.8.

Where the failure to achieve agreement with the subcontractor within ten days results in the notice required by clause 35.8.2 (**document 8.07.2**), the architect has the option of either refusing to accept your reasons for non compliance or accepting them. In the former case he will require you to execute Agreement NSC/A with the subcontractor without any further instructions from him. The contract is silent as to what you should do if you still cannot reach agreement with the subcontractor. Your only option appears to be to issue a further notice under clause 35.8.2 explaining your problem. Where the architect accepts your explanation he is required to issue further instructions, cancel the nomination instruction or nominate another sub-contractor.

There used, in JCT 80, to be an alternative and simpler method but this no longer forms a part of the contract.

8.07.1 Letter regarding the failure to complete negotiations with a nominated subcontractor within ten days

---

To the Architect   Date

Dear Sir,

[*Heading*]

We refer to your instruction nominating Hardasiron Steelwork Ltd as nominated subcontractors for the structural steelwork on the Contract, received on [*date*]. In accordance with Clause 35.8.1 of the Conditions of Contract, we have to inform you that we have been unable to reach agreement in respect of form NSC/T Part 3 and have not therefore been able to execute Agreement NSC/A within 10 working days of receipt of your notice.

We anticipate that we shall be able to complete our negotiations by [*date*] and we therefore ask you to set that date as the revised date for compliance with your instruction in accordance with Clause 35.9.1.

Yours faithfully,

8.07.2   Letter to the architect where you are unable to reach agreement with a proposed nominated subcontractor

---

To the Architect   *Date*

Dear Sir,

[*Heading*]

We refer to your instruction nominating Hardasiron Steelwork
Ltd as nominated subcontractors for the structural steelwork
on this contract, received by us on [*date*].
   In accordance with Clause 35.8.2 of the Conditions of
Contract we have to inform you that we have been unable to
complete these negotiations within the time specified because
[*state reason, e.g.* Hardasiron Steelwork Ltd require a period
for fabrication which would mean that delivery of steelwork to
the site could not commence until three weeks after the date
specified in our master programme and will require a further
payment of £[*x*] if delivery is to be accelerated to accord with
our programme].

We await your instructions under Clause 35.9.2 of the
Contract. [In the meantime we are continuing with
negotiations (but we must point out that unless negotiations
are completed and a subcontract entered into within 7 days from
today it will be impossible for Hardasiron Steelwork Ltd to
effect delivery by our programmed date even if the extra
payment which they require is agreed to).]

Yours faithfully,

8.08   Right of objection

Clause 35.5.1 of JCT 98 gives you an important safeguard. It says 'No person against whom the Contractor makes a *reasonable* objection shall be a Nominated Sub-Contractor'. If the architect rejects your objection it is a dispute that is referable to adjudication. There is no guidance in the contract as to what is a reasonable objection. You cannot object, however, to details in the specification of the subcontract or the subcontract price but the following would be reasonable grounds for objection:

- The subcontractor's programme will not fit in with your programme.
- The subcontractor is financially unstable – rumour is insufficient, you need hard facts.
- The subcontractor is too large or too small.
- The subcontractor is too inexperienced.
- Your past experiences with him have shown him to be unreliable, etc.

You must make your objection 'at the earliest practicable moment' but in any case 'not later than 7 working days from receipt of the instruction of the Architect under Clause 35.6 nominating the subcontractor'.

Time, therefore, is of the essence. A telephone objection is valid, but it is best to put your objection in writing. Be sure of your facts and that your objection is reasonable (**document 8.08.1**).

8.08.1 Objection to a nomination

To the Architect　　　　　　　　　　　　　　　　　　　　　*Date*

Dear Sir,

[*Heading*]

We are in receipt of your instruction nominating Hardasiron
Steelwork Ltd as subcontractors for the structural steelwork
on this Contract.

We regret to inform you that we have a reasonable objection to
this intended nomination in that [*specify reason(s), see text
for suggested list*].

In accordance with Clause 35.5.2 of the Conditions of the
Contract we shall be grateful, as a matter of urgency, [to
receive your instruction that will remove the objection so
that we can comply with Clause 35.7 *or* if you would either
cancel your nomination instruction and issue an instruction
omitting the work or nominate another subcontractor].

Yours faithfully,

8.09   Procedures after nomination

Once a subcontract is in being, there are many letters and notices which you may need to send. Clause 35 of JCT 98 covers a variety of situations, and the sending of some notices etc. is mandatory. In other cases it is wise to get something in writing.

Payment

Clause 35.13 deals with payment of the nominated subcontractor and if you fail to pay the nominated subcontractor there are provisions for direct payment.

You are required by clause 4.16.1.1 of NSC/C to give a written notice to the subcontractor not later than five days after the date of issue of an interim certificate. This written notice must show two things. Firstly, in respect of the amount set out in the certificate (Amount A) you must show to what the amount relates and the basis on which it was calculated. Secondly, you must show 'Amount B', which is any other amount due to the subcontractor under the subcontract which has to be supported by similar information to that required for Amount A. **Document 8.9.1** is an example of the covering letter to this notice.

After the first interim certificate, before the issue of each subsequent certificate, it is your duty to provide the architect 'with reasonable proof' that you have paid the nominated subcontractor. The nominated subcontractor is required to provide you with written proof that you can use to provide the architect with the reasonable proof of payment that he in turn requires. **Document 8.9.2** is a suitable letter for you to send to the subcontractor enclosing payment and requiring receipt.

If the subcontractor fails to respond then, before the next certificate falls due, you should write to the architect explaining the situation (**document 8.9.3**) which explains why you are unable to provide the reasonable proof required.

The architect cannot operate the direct payment provisions unless he has issued a certificate that you have failed to provide reasonable proof that the payment to the subcontractor has been made. Sometimes such certificates are issued unreasonably and in that case you must protest: see **document 8.9.4**.

You may consider that you are entitled to reduce the amount of the payment set out by the architect in his certificate because of some default of the subcontractor. Clause 4.16.1.2 sets out the procedure that you must follow. This requires that you issue a written notice specifying the amount proposed to be withheld and/or deducted from the amount otherwise payable. This must be done no later than five days before the final date for payment (17 days from the date of issue of the interim certificate in which the amount is included).

Failure to issue a withholding notice before paying less than the amount set out in the notice required by clause 4.16.1.1 may result in the subcontractor issuing a seven day notice of suspension of performance or commencing adjudication proceedings unless the amount is paid in full.

Extension of time to nominated subcontractor

You cannot grant any extension of time to the nominated subcontractor unless the architect gives his prior written consent. Any extension granted must be in accordance with clauses 2.2–2.6 of the subcontract form (see section 8.12). **Document 8.9.5** is a letter which may be used when you pass on to the architect your nominated subcontractor's application for extension of time and/or the supporting particulars and estimate.

Failure to complete

If the nominated subcontractor fails to complete the subcontract works within the period specified or as extended by you, you must notify the architect immediately (**document 8.9.6**), sending a copy to the defaulting subcontractor. The architect must then issue a certificate of default to you, with a duplicate to the nominated subcontractor. You are advised to use special delivery post.

You should note that the architect must issue this certificate of default – provided he is satisfied that the extension of time provisions have been properly operated – within two months from the date on which you notify him of the subcontractor's default. If he fails to do so – or if he disputes that the nominated subcontractor is in default – you must protect your own interests by writing to the architect immediately (**document 8.9.7**).

Defects liability

The contract gives the nominated subcontractor valuable rights to early final payment provided certain conditions are satisfied. Problems can arise if, having been paid, the nominated subcontractor refuses to rectify defects, shrinkages or other faults in the subcontract works which he is bound to rectify. In that event, clause 35.18 enables the architect to issue you with an instruction nominating a substituted subcontractor to carry out the remedial works. You must be given the opportunity to agree the substituted subcontractor's price before the nomination is made. **Document 8.9.8** is a letter which is freely adaptable in these circumstances.

Limitation of liability

In the case of nominated subcontractors you are not responsible to the employer for defects in (a) design, (b) selection of materials and goods, (c) the satisfaction of any performance specification or requirement and (d) the provision of any information required to be provided pursuant to Agreement NSC/W. These are the nominated subcontractor's responsibility, and you have no liability to the employer in these matters, even if the employer has not taken the precaution of making the nominated subcontractor enter into a warranty agreement in terms of agreement NSC/W. This position contrasts with your liability for domestic subcontractors and, of course, you remain fully responsible to the employer for any other defaults by a nominated sub-contractor.

Some architects, apparently, misunderstand this position and through inadvertence or otherwise, may attempt to fix you with liability. Any attempt of this sort should be resisted strongly (see **document 8.9.9**).

8.9.1 Notice under clause 4.16.1.1 of NSC/C

To the Subcontractor   *Date*

Dear Sirs,

[*Heading*]

In accordance with Clause 4.16.1.1 we enclose herewith our notice setting out the amount stated as due to you in the Architect's Interim Certificate No. [*x*]. We have also, as required, set out to what the amount relates and the basis on which it is calculated.

[We reserve our right to issue the notice envisaged by Clause 4.16.1.2 in respect of any sums that are withheld from the payment.]

This sum, subject to any amount withheld, is due to you no later than [*date*], which is the final date for payment.

Yours faithfully,

8.9.2 Payment to a nominated subcontractor

To the Subcontractor *Date*

Dear Sir,

In accordance with Clause 35.13.2 of the Conditions of the main Contract and Clause 4.16.1.1 of our subcontract, we enclose our cheque for [£x] as payment to you of the sum directed by the Architect in the statement attached to the Interim Certificate No. [x] and as set out in our notice to you of [date]. Please sign and return the attached receipt <u>within 7 days from the date of this letter</u> as the proof of discharge required by Clause 4.16.1.1 of our subcontract.

Yours faithfully,

8.9.3 A nominated subcontractor's failure to provide proof of discharge

---

To the Architect   *Date*

Dear Sir,

[*Heading*]

With reference to your direction regarding payment to
Hardasiron Steelwork Ltd contained in the Statement attached
to your Interim Certificate No. [*specify number*] we duly
discharged that payment in accordance with Clause 35.13.2 of
the Conditions of the main Contract and Clause 4.16.1 of the
subcontract. We enclose a copy of our letter to the firm and of
our cheque for the amount shown in your direction less $2\frac{1}{2}\%$ cash
discount as set out in the attached notice.

However, Hardasiron Steelwork Ltd have failed to return the
receipt as requested in our letter. Since they have failed to
provide the 'document or other evidence' referred to in Clause
35.13.4 of the Conditions of the main Contract, we trust that
you will accept the attached copy letter and cheque as the
'reasonable proof of payment' required by Clause 35.13.3 of
the Conditions.

Yours faithfully,

8.9.4   Letter protesting at the issue of a certificate of non-payment

SPECIAL DELIVERY

To the Architect   *Date*
(Copy to the Employer)

Dear Sir,

[Heading]

We note that you have issued to the Employer a Certificate
under Clause 35.13.5.1 of the Conditions of the main Contract
stating that we have failed to provide 'reasonable proof of
payment' of the payment directed to be made to Hardasiron
Steelwork Ltd in the Statement attached to your Interim
Certificate No. [*specify number*].

As stated in our letter to you of [*date*], Hardasiron Steelwork
Ltd failed to return the receipt attached to our letter
enclosing payment. In those circumstances the copy letter and
cheque enclosed with our letter to you of [*date*] informing you
of Hardasiron Steelwork Ltd's failure to return our receipt
should have been acceptable to you as the 'reasonable proof of
payment' required. The issue of your Certificate under Clause
35.13.5.1 of the Contract Conditions is therefore
unreasonable, and we must request you to withdraw it
forthwith.

Should the Employer make direct payment to the subcontractor
in reliance on that Certificate and deduct the amount paid from
payment under interim Certificate No. [*specify number*]
recently issued we shall issue an immediate Notice of
Adjudication.

Yours faithfully,

8.9.5 Letter passing on a notice of delay etc. from a nominated subcontractor

---

To the Architect   *Date*

Dear Sir,

[*Heading*]

In accordance with clauses 25.2 and 35.14 of the Conditions of the main Contract and Clause 2.2 of our subcontract with Hardasiron Steelwork Ltd, we enclose herewith a notice of delay from that firm together with supporting particulars and estimates as required by Clause 2.2.3 of the subcontract which we have received today. [*Or* we enclose particulars and estimates which we have received today relating to the notice of delay submitted by that firm and forwarded to you under cover of our letter dated [*date*].]

In accordance with Clause 35.14.2 of the Conditions of the main Contract and Clause 2.3 of the subcontract, we shall be glad to receive your written consent so that we may grant any appropriate extension of time to the subcontractor within the time limit of 12 weeks from today specified in Clause 2.3 of the subcontract. Please let us know if you require any further particulars or estimates from the subcontractor to enable you to reach a decision as to the granting of consent.

[*If appropriate you could add at the end of the first paragraph:* You will note that this notice refers to the delays referred to in our notice of delay under Clause 25.2 of the Conditions of the main Contract, which accompanies this letter.]

Yours faithfully,

8.9.6 Letter notifying the architect of a nominated subcontractor's failure to complete

SPECIAL DELIVERY

To the Architect   *Date*
(Copy to the Subcontractor)

Dear Sir,

[*Heading*]

In accordance with Clause 35.15.1 of the Conditions of the main
Contract, this is to notify you that Hardasiron Steelwork Ltd,
as nominated subcontractors for the steelwork on this
Contract, have failed to complete their subcontract works
within the period specified in their subcontract with us [*or*
within the extended period granted by us in accordance with
your written consent given in your letter dated [*date*]],
that period having expired yesterday.

We shall be grateful to receive your Certificate to that effect
as soon as possible so that we may avail ourselves of the
remedies provided under the subcontract in relation to the
effects of this subcontractor's default.

Yours faithfully,

8.9.7   Letter regarding the architect's refusal to grant a certificate of the nominated subcontractor's delay

To the Architect   *Date*
(Copy to the Subcontractor)

Dear Sir,

[Heading]

With reference to our letter of *[date]* notifying you under Clause 35.15.1 that Hardasiron Steelwork Ltd had failed to complete their subcontract works within the period specified in their subcontract *[Or* within the extended period for completion of their works granted by us with your written consent]*, the period of 2 months allowed for the issue of your Certificate to that effect has now expired and we note from your letter dated *[date]* that you are refusing to issue the Certificate on the grounds that we have failed properly to apply the terms of Clause 35.14 of the Conditions of Contract.

We must draw your attention to our letter of *[date]* with which we enclosed the subcontractor's notice of delay dated *[date]* together with the particulars and estimates provided by them, all in accordance with our obligation under Clause 35.14 of the Conditions of the main Contract and under Clause 2.2 of the subcontract. We have received no other notices of delay from this subcontractor. We have therefore fulfilled our obligation under Clause 35.14 of the Conditions of Contract and we do not consider that you have any valid reason for your refusal to grant the required Certificate under Clause 35.15 of the Conditions.

Will you please, therefore, issue your Certificate forthwith, failing which we shall have no alternative but to notify the Employer under Clause 41 of the Contract that a dispute exists and issue an immediate Notice of Adjudication.

Yours faithfully,

8.9.8 Letter regarding the agreement of a price submitted for the rectification of defects

---

To the Architect   *Date*

Dear Sir,

[*Heading*]

Thank you for your letter of [*date*] enclosing a quotation submitted by Acme Heating Engineers (1989) Ltd for the rectification of defects in the heating installation which the original Nominated Subcontractor has refused to rectify. We agree to the price quoted and therefore consent to the nomination of this firm as 'substituted subcontractor' under Clause 35.18.1.1 of the Conditions of Contract.

We confirm that the 'substituted subcontractor' will be subject to all the provisions relating to nominated subcontractors in Clause 35 of the Conditions of Contract.

[*Or* We regret that we cannot agree to the price quoted and must therefore refuse to enter into a subcontract with this firm as 'substituted subcontractor'. We enclose a quotation which we have received for this work from VPS Engineering Ltd which you will note is substantially lower than that enclosed with your letter. The firm is well known to us and is of high repute, and we shall therefore be glad if you will issue your instruction nominating VPS Engineering Ltd as substituted subcontractor under Clause 35.18.1.1 of the Conditions of Contract.]

Yours faithfully,

8.9.9 Letter regarding liability for a nominated subcontractor's design failure

---

To the Architect *Date*

Dear Sir,

[*Heading*]

We are in receipt of your letter of [*date*] regarding the failure of the heating installation executed by Acme Heating Engineers (1989) Ltd to achieve the heat output required by the specification, and suggesting that the Employer may seek to proceed against us in respect of this failure.

This failure is clearly due to failure by Acme to exercise reasonable skill and care in the design of the system and in the satisfaction of the performance specification. As such it is of the class of failure with regard to which any responsibility to the Employer on our part is expressly excluded by Clause 35.21 of the Conditions of the main Contract. [This exclusion is not affected by the fact that the nomination of this subcontractor was made under Clause 35.11 of the Conditions and the Employer has chosen not to enter into an Agreement with the subcontractor under Form NSC/W.] The Employer therefore has no cause of action against us in respect of this failure.

Yours faithfully,

8.10 Re-nomination

Where the subcontractor 'drops out' the contract lays down a clear procedure which you must follow in order to get a re-nomination. You have no right or duty to do the nominated subcontract works yourself and the architect must nominate another firm. As main contractor you have to carry the *immediate* financial and other consequences of a nominated subcontractor's failure or withdrawal – whatever the reason – and the architect has a 'reasonable time' in which to make the re-nomination.

This period of time does not begin to run until you have made a formal application to him for a fresh nomination instruction. You are only entitled to claim an extension of time (clause 25.4.7) and loss and/or expense (clause 26.4) if the architect fails to re-nominate within a reasonable time. A period of several weeks from the date of your application for a fresh nomination instruction may well be a 'reasonable time' because the architect has to seek out fresh tenders and so on.

The re-nomination procedure is laid down in clause 35.24, which covers five situations:

- Default by nominated subcontractor under the terms of the subcontract.
- Insolvency of nominated subcontractor.
- Valid determination by the nominated subcontractor of his own employment under the subcontract.
- You have been required by the employer to determine the employment of the nominated subcontractor.
- Where work executed by the nominated subcontractor has to be taken down and the nominated subcontractor cannot be required under the terms of the nominated subcontract or does not agree to carry out works, to take down and/or re-execute or re-fix or re-supply work properly executed or materials or goods properly fixed or supplied.

Default by the nominated subcontractor

Clause 7.1.1 of NSC/C specifies the defaults by the nominated subcontractor which may entitle you to determine his employment if you go through the correct procedure. These defaults are:

- If the subcontractor *wholly or substantially* suspends the carrying out of the subcontract works before completion, without reasonable cause. (The important word here is 'suspends'. 'Going-slow' is not suspending the works and you will have to be very careful before deciding to operate this provision if the nominated subcontractor is still working on site.)
- If the subcontractor fails to proceed regularly and diligently with the subcontract works, again without reasonable cause.

- If the subcontractor refuses or persistently neglects to remove work materials or goods not in accordance with the contract after due notice.
- If the subcontractor assigns or sublets without consent.
- If the subcontractor fails to comply with the requirements of the CDM Regulations.

If you are sure that the subcontractor is at fault in one or more of these respects you must inform the architect in writing (**document 8.10.1**) and send to him any written observations which the subcontractor has in regard to the alleged default(s). If the architect is reasonably of the opinion that the subcontractor has made default, he must issue you with an instruction to give the subcontractor notice of default. A specimen notice of default is shown as **document 8.10.2**. It must be sent by actual, special or recorded delivery, with a copy to the architect.

If the subcontractor continues the default for 14 days after receipt of your notice and you wish to determine his employment, you should send a letter (**document 8.10.3**) within ten days by actual, special or recorded delivery to determine forthwith the subcontractor's employment. However, the architect may issue a further instruction under clause 35.24.6.1 of the main contract before you exercise this right. In that case you must request his permission to so determine the employment of the nominated subcontractor (**document 8.10.4**). The time limit is very tight.

You should note that a repetition of the default at any time after the original notice entitles you to determine forthwith without a further 14 days notice, but in this case the architect once again has the option to issue a further instruction. Because the consequences of a *wrongful* determination are so serious from every point of view, regard this procedure as the last resort when all else has failed.

An immediate reference to adjudication in an attempt to resolve any underlying dispute that is causing the default of the nominated subcontractor may be appropriate.

Whether or not you have determined the subcontractor's employment, you must inform the architect (**documents 8.10.5** and **8.10.6**) and we suggest that you ask for a re-nomination which the architect is bound to make. If you have determined the subcontractor's employment on the third ground (failure to remedy defective work etc.) you must be given an opportunity to agree the price to be charged by the substituted subcontractor, but you cannot withhold your agreement unreasonably: see **document 8.9.8** in previous section.

Insolvency of nominated subcontractor

The provisions of both JCT 98 and NSC/C are intricate where a nominated subcontractor becomes insolvent.

The aspects of insolvency which are set out in NSC/C in clause 7.2.1 are that the subcontractor

- makes a composition or arrangement with his creditors or becomes bankrupt

or being a company

- makes a proposal for a voluntary arrangement for a composition of debts or scheme arrangement
- has a provisional liquidator appointed
- has a winding up order made
- passes a resolution for voluntary winding up
- has an administrator or administrative receiver appointed

You have an obligation to inform the architect when any of the insolvency events in clause 7.2.1 of the subcontract occurs.

In a number of situations, for example where a provisional liquidator or trustee in bankruptcy is appointed (see clause 7.2.3 of NSC/C), the employment of the subcontractor is automatically determined. All you need to do is to inform the architect accordingly and seek a re-nomination. **Document 8.10.7** is a suitable letter to write in these circumstances. You may, however, with the written consent of the architect and the agreement of the subcontractor, reinstate his employment. You have no obligation to do this and would be very well advised only to do this where it is clearly to your advantage. Note also that clause 7.2.3 of the subcontract is not one of those clauses that puts an obligation on the architect to give his consent in all but unreasonable circumstances.

In any insolvency situation other than those described in clause 7.2.3 of the subcontract the permission of the architect is required before you can determine the subcontract. **Document 8.10.8** is the sort of letter to write then.

Determination of the nominated subcontractor's employment on the instruction of the employer

An instruction of this nature will only be given where the nominated subcontractor or any person employed by or acting on his behalf (with or without his knowledge) transgresses the requirements of clause 7.3 of the subcontract which deals with corruption. You will be entitled to a re-nomination by the architect in this instance.

Determination by the subcontractor

You are in trouble if the nominated subcontractor validly determines his own employment under the subcontract. He can do this under clause 7.7 of the subcontract on three grounds:

- If you wholly suspend the main contract works before completion without reasonable cause.
- If you fail to proceed with the main contract works without reasonable cause so that the progress of the subcontract works is seriously affected.
- If you fail to comply with the CDM Regulations.

The subcontractor must go through the proper procedure but, assuming that he has done so and has validly determined his own employment, the architect must re-nominate as necessary, and you may find yourself having to meet the extra costs.

If a nominated subcontractor threatens determination, you should take competent legal advice at once, as if he is successful the results will be disastrous for you.

Finally, there is a final important point about determination. You are not entitled to determine the subcontractor's employment under the contract or to terminate the subcontract for breach of contract at common law unless you obtain an architect's instruction to do so. The legal textbooks discuss the circumstances in which the subcontractor's conduct amounts to a common law breach of contract. Legal advice is always necessary but, if you are in this unfortunate situation, you must remember to seek an instruction from the architect (**document 8.10.9**).

Where work is to be re-executed

You may find yourself in the position that work that has already been properly executed by a nominated subcontractor needs to be taken down and re-executed in order that you can comply with an architect's instruction. Clause 35.24.5 of YCT 98 covers the situation where you are unable to insist that the nominated subcontractor does the taking down and re-execution of the work and he refuses to do so.

8.10.1 Letter informing the architect of a nominated subcontractor's default

---

To the Architect   *Date*

Dear Sir,

[*Heading*]

In accordance with Clause 35.24.1 of the Conditions of
Contract, this is to inform you that, in our opinion, the
Nominated Subcontractor for the heating installation, Acme
Heating Engineers (1989) Ltd, is in default under Clause
7.1.1.2 of the subcontract in that the firm without reasonable
cause is failing to proceed regularly and diligently with the
subcontract Works.

[We enclose a letter from the subcontractor dated [*date*]
setting out some observations in relation to this default. In
our view those observations do not disclose any reasonable
cause for the firm's default.]

If you are of the opinion that the subcontractor is in default
in the manner which we allege, please issue your instruction to
us to issue the notice of default specified in Clause 7.1.1 of
the subcontract. Please let us know if there is any further
information which you require in order to enable you to form
your opinion on the matter. In accordance with Clause
35.24.6.1 of the Conditions of Contract, please also inform
us, when issuing your instruction, if you will require us to
obtain a further instruction from you before determining the
subcontractor's employment should they not rectify their
default following our notice.

Yours faithfully,

8.10.2   Notice of a nominated subcontractor's default

SPECIAL DELIVERY

To the Nominated Subcontractor   *Date*
(Copy to the Architect)

Dear Sirs,

[Heading]

As instructed by the Architect under Clause 35.24.6.1 of the
Conditions of the main Contract, we hereby notify you that you
are in default under Clause 7.1.1.2 of the subcontract in that,
without reasonable cause, you are failing to proceed
regularly and diligently with the Works.

Should you continue with this default for a period of 14 days
following receipt of this notice, or at any time subsequently
repeat this default, we shall [if so instructed by the
Architect]* determine your employment under the subcontract
as provided in Clause 7.1 thereof.

Yours faithfully,

*Omit words in brackets if the Architect has not said that you
must obtain a further instruction before determination.*

8.10.3 Notice determining the nominated subcontractor's employment

---

SPECIAL DELIVERY

To the Nominated Subcontractor   *Date*
(Copy to the Architect)

Dear Sirs,

[Heading]

With reference to our notice of default dated [*date*], which we
have been notified by Royal Mail was received by you on [*date*],
you have continued the specified default for a period of 14
days following receipt thereof.

Accordingly [as instructed by the Architect]* we hereby
forthwith determine your employment under the subcontract.

We must point out that, by Clause 7.5.1 of the subcontract, any
person nominated by the Architect to carry out and complete the
subcontract Works may enter thereon and use all temporary
buildings, plant, tools, equipment, goods and materials
intended for, delivered to and placed upon or adjacent to the
Works. We shall communicate with you further in due course
concerning the matters set out in Clauses 7.5.2 to .4 of the
subcontract.

Yours faithfully,

*Omit words in brackets if the architect has not said that you
must obtain a further instruction before issuing this notice.*

8.10.4 Letter requesting the architect's instruction to determine the nominated subcontractor's employment

---

SPECIAL DELIVERY

To the Architect   *Date*

Dear Sir,

[Heading]

Following our notice of default dated *[date]*, a copy of which was sent to you, which we have been informed by the Royal Mail was delivered to Acme Heating Engineers (1989) Ltd on *[date]*, the firm has continued with the default for the period of 14 days specified in Clause 7.1.2 of the subcontract which expires today.

Will you please therefore instruct us to determine the firm's employment under the subcontract. We shall be grateful if you will issue that instruction immediately so that we may issue the notice of determination within the period of 10 days from today specified in Clause 7.1.2 of the subcontract.

Yours faithfully,

8.10.5   Letter notifying the architect of the determination of the nominated subcontractor's employment

To the Architect   *Date*

Dear Sir,

[*Heading*]

Following our notice of default dated [*date*] a copy of which
was sent to you [*or* Following your Architect's instruction No.
595 dated [*date*], we now enclose the notice which we have sent
today to Acme Heating Engineers (1989) Ltd determining their
employment under their subcontract.

As provided in Clause 35.24.6.3 of the Conditions of the main
Contract, please make further nomination of a subcontractor
to carry out and complete the remaining subcontract Works as
soon as possible so that delay in continuing with the Contract
Works generally may be kept to the minimum. Please also let us
know of any requirement which you or the Employer may have for
the assignment to us of the benefit of any agreement for the
supply of materials or goods and/or for the execution of any
work for the purposes of the subcontract and let us have any
directions as to payment of any supplier or sub-subcontractor
as provided in Clause 7.5.2.2 of the subcontract. We shall
telephone you within the next day or two to arrange a meeting to
discuss these and any other matters arising from the
determination of this Nominated Subcontractor's employment.

Yours faithfully,

8.10.6   Letter to the architect informing him of the decision not to determine the nominated subcontractor's employment

To the Architect   *Date*

Dear Sir,

[*Heading*]

Following our notice of default dated [*date*] addressed to Acme
Heating Engineers (1989) Ltd, a copy of which was sent to you,
the firm has now rectified the default specified in the notice.
We have therefore decided not to proceed with the issue of a
notice of determination [*or* not to seek your further
instruction to issue a notice of determination].*

Clause 7.1.3 of the subcontract entitles us to determine this
Nominated Subcontractor's employment forthwith should they
at any time repeat the default of which they were notified.
Should this situation arise we shall immediately notify you
[and request your instruction then to determine the firm's
employment].*

Yours faithfully,

*Omit words in brackets if the Architect has not said that you
must obtain a further instruction before determination.*

8.10.7 Letter to the architect notifying him of the nominated subcontractor's insolvency (clause 7.2.3 of NSC/C)

To the Architect   *Date*

Dear Sir,

[Heading]

We have just received notification that a provisional liquidator has been appointed to wind up the affairs of Acme Heating Engineers (1989) Ltd, the nominated subcontractor for the heating and ventilation services on this Contract. By Clause 27.2 of the subcontract the employment of this firm as nominated subcontractor has therefore been automatically determined.

As provide in Clause 35.24.7.3 of the Conditions of the main Contract, please make further nomination of a subcontractor to carry out and complete the remaining subcontract works as soon as possible so that delay in continuing with the Contract Works generally may be kept to the minimum. We shall telephone you in the next day or two to arrange a meeting to discuss matters arising from the determination of this nominated subcontractor's employment.

Yours faithfully,

8.10.8 Letter to the architect seeking consent to determine nominated subcontractor's employment (clause 7.2.4 of NSC/C)

To the Architect *Date*

Dear Sir,

[*Heading*]

We have just been informed by Acme Heating Engineers (1989) Ltd, the Nominated Subcontractor for the heating and ventilation services on this Contract, that they are making proposals for a voluntary arrangement for a composition of debts.

We are very concerned that this will make their already slow progress even slower with the consequent effect on the overall completion of the project.

The only sensible way forward in our view is to determine Acme Heating Engineers' employment under the subcontract.

We therefore seek the written consent from you required by clause 7.2.4 of the subcontract and seek a re-nomination under Clause 35.25.7.3. of the main Contract.

Yours faithfully,

8.10.9 Letter to the architect seeking instruction to determine a subcontract

To the Architect *Date*

Dear Sir,

[*Heading*]

We notified you in our letter of [*date*] that, in our opinion, Acme Heating Engineers (1989) Ltd are in fundamental breach of their obligations under the subcontract with us as nominated subcontractor for the heating and ventilation services on this Contract. This opinion has now been confirmed by legal advice, a copy of which is enclosed. We therefore now regard the firm as having repudiated their subcontract with us.

As required by Clause 35.25 of the Conditions of the main Contract, please issue your instruction to us to determine our subcontract with Acme Heating Engineers (1989) Ltd.

Will you please also take steps to nominate a further subcontractor to carry out and complete the remaining subcontract Works as soon as possible so that delay in continuing with the Contract Works generally may be kept to a minimum. We shall telephone you in the next day or two to arrange a meeting to discuss matters arising from the determination of this subcontract.

Yours faithfully,

8.11   Named subcontractors under IFC 98

IFC 98 contains no provisions for the 'nomination' of subcontractors or suppliers. All suppliers, even if specified in the contract documents, will be ordinary 'domestic' suppliers for whom you have full responsibility. However, IFC 98, clause 3.3, contains complex provisions enabling the employer to select a named subcontractor to carry out part of the work and you must employ the person so named.

There are two methods of selection. Subcontractors can be named in the original contract documents, or can be named by the architect/contract administrator when issuing an instruction for the expenditure of a provisional sum as the contract progresses. The two procedures for naming a subcontractor are given here.

Procedure (1)

The employer provides a detailed description of the work in the contract documents. The work is to be priced by you and is to be carried out by the named person. There is a standard form of tender and agreement (NAM/T), section I of which is the invitation to tender and contains relevant main contract information, expected commencement dates and subcontract periods. Section II is the subcontractor's tender in which he also sets out his requirements. You have no right of reasonable objection to the person named under this procedure since you know his identity from the outset.

Under this procedure you must enter into a subcontract with the named person within 21 days of the main contract using section III of NAM/T which incorporates the standard subcontract NAM/SC.

This is a very tight timescale and you will have tendered on the basis of the particulars given in the contract documents relating to the subcontract works. If you cannot agree the particulars with the subcontractor within 21 days you must inform the architect immediately, detailing the particulars which have prevented the subcontract being made (**document 8.11.1**).

If the architect is reasonably satisfied that the particulars specified by you are causing the problem he can then:

- change the particulars so as to remove the impediment, *or*
- omit the work altogether, *or*
- substitute a provisional sum – against which, of course, he can then name a new subcontractor.

The architect can take the last of these courses in any case even where there is no problem – but not, of course, after you have actually entered into a subcontract with the original firm.

Where the architect changes the particulars or omits the work the instruction is treated as a variation; the contract sum is adjusted and you are entitled to any appropriate extension of time and 'direct loss and/or expense'. If the work is omitted the employer can have it done by somebody else for whom he will then take full responsibility.

Procedure (2)

This procedure applies where the subcontractor is named in a provisional sum instruction. This must provide the detailed information contained in sections I and II of NAM/T, and you have a right to make a reasonable objection to the person named. You must make your objection by writing to the architect (**document 8.11.2**) within 14 days of the date of *issue* (not receipt) of the instruction. Any objection may be submitted to adjudication by the employer as a dispute. Assuming you have no objection to the person named in the provisional sum instruction, you must enter into a subcontract with him using section III of NAM/T within 14 days of the *issue* of the instruction. This is a very short period and really gives you insufficient time; if any of the particulars are not acceptable and you cannot settle them with the subcontractor within the 14 day period you should notify the architect and **document 8.11.1** may be adapted. If compliance with the architect's instruction directly involves you in delay or disturbance to contract progress – perhaps because of programming difficulties – then you are entitled to both an extension of time and to direct loss and/or expense.

General

If it becomes likely that the subcontractor's employment will have to be determined you must notify the architect (**document 8.11.3**). You can only determine the subcontractor's employment under clauses 27.2 to 27.4 of NAM/SC and you must not let the employment be determined by accepting a repudiation by the named subcontractor of the subcontract. Whether you give such notice or not, if the subcontractor's employment actually is determined you must again notify the architect telling him why (**document 8.11.4**). The architect then may:

- name another subcontractor, *or*
- instruct you to do the work, in which case you can sublet it to someone of your choice, with consent (**document 8.11.5**), *or*
- omit the work from the contract, in which case the employer can get it done by someone else.

If the architect takes the first course the contract sum will be adjusted to take account of the changed price of the new subcontractor. You will be entitled to any appropriate extension of time (but no 'loss and/or expense') but you will be responsible for the cost of making good defects in the first subcontractor's work. You have the opportunity to approve the price charged by the new subcontractor for making such defects good. If the architect takes either of the other two courses this will be treated as a variation; the contract sum will be adjusted and you will be entitled to any appropriate extension of time and 'direct loss and/or expense'. However, there will be no increase in the contract sum, no extension of time and no 'loss and/or expense' if the determination is your fault.

Clause 3.3.6(b) puts you under an onerous obligation to pursue the defaulting subcontractor for the extra costs incurred by the employer following the architect's instruction where you have properly determined the subcontractor's employment. You must try to recover any additional amounts payable to you by the employer as a consequence of the determination, together with an amount equal to any liquidated damages which the employer would have recovered had the extension of time provisions not been operated, and you must account to the employer for any sums recovered (**document 8.11.6**). If you fail to comply with these provisions you are liable to the employer for any extra costs and an amount equal to the lost liquidated damages. However, you are not bound to start litigation, arbitration or adjudication proceedings unless the employer provides you with an indemnity against reasonable legal costs (**document 8.11.7**).

8.11.1 Letter to architect where unable to enter into subcontract with named person in accordance with particulars (IFC 98)

To the Architect *Date*

Dear Sir,

[*Heading*]

In accordance with Clause 3.3.1 of the Contract we must inform you that we are unable to enter into a subcontract with [*name*] in accordance with the particulars given in the Contract Documents.

The particulars which prevent the execution of the subcontract are: [*give details*].

Will you please let us have the necessary instruction.

Yours faithfully,

8.11.2   Letter making reasonable objection to proposed named subcontractor (IFC 98)

To the Architect   *Date*

Dear Sir,

[*Heading*]

We are in receipt of your instruction No. [*insert number*] dated [*insert date*] issued under Clause 3.3.2(a) requiring us to enter into a subcontract with [*X*] as named subcontractors.

We regret to inform you that we have a reasonable objection to this named person in that [*specify reasons*].

We shall be glad to receive your instructions in this matter.

Yours faithfully,

8.11.3 Letter to architect giving notice of likely determination of named subcontractor's employment (IFC 98)

To the Architect   *Date*

Dear Sir,

[*Heading*]

In accordance with Clause 3.3.3 of the Contract we must advise you that the following events are likely to lead to the determination of the employment of [*insert name*] under Clause 27.1 of the Subcontract.

[*Insert details of events*]

May we please have your instructions.

Yours faithfully,

8.11.4 Letter to architect informing him of determination of named subcontractor's employment (IFC 98)

To the Architect   *Date*

Dear Sir,

[*Heading*]

In accordance with Clause 3.3.3 of the Contract, we must inform you that the employment of [*insert name*] was determined on [*date*] under [*specify subcontract clause or other circumstances*] and we shall be pleased to receive your immediate instructions.

[*If you have not already advised the Architect of the likelihood of determination, add:* Because of the nature of the events leading to this determination, we regret that we were unable to give you prior notice of its likely happening and trust that you will understand.]

Yours faithfully,

8.11.5 Letter to architect where contractor instructed to carry out work after determination and seeking consent to subletting (IFC 98)

To the Architect   *Date*

Dear Sir,

[*Heading*]

Further to your Instruction No. [*insert number*] of [*date*] under Clause 3.3.3(b) of the Contract requiring us to make our arrangements for the execution of the outstanding work, we are proposing to make arrangements with [*insert name*] who specialise in this type of work and we shall be grateful if you will consent to our subcontracting with them.

Yours faithfully,

8.11.6   Letter to architect regarding recovery of costs from named subcontractor after determination (IFC 98)

To the Architect   *Date*

Dear Sir,

[*Heading*]

In accordance with Clause 3.3.6(b) of the Contract, we have taken reasonable steps under Subcontract Clause 27.3.3 to recover the employer's extra costs and lost liquidated damages.

The steps we have taken are:
[*Give details, and enclose supporting documentation*]

[We are pleased to say that we have today received a cheque for £x from [*insert name*] made up as follows [*give details*] and we enclose our cheque for that amount.]

[*If efforts are wholly or partly unsuccessful, give explanation and add:* In the circumstances, short of initiating arbitration or litigation it appears to us that there are no further reasonable steps which we can take. We are, of course, prepared to take proceedings if the employer will indemnify us against the necessary legal costs of so doing.]

Yours faithfully,

8.11.7 Letter seeking employer's indemnity against legal costs (IFC 98)

To the Architect *Date*

Dear Sir,

[*Heading*]

Thank you for your letter of [*date*] informing us of the Employer's wish that we should commence arbitration proceedings against [*insert name*] in an attempt to recover the additional amounts payable to us by him and his lost liquidated damages consequent on the determination of [*X's*] employment.

We are prepared to commence proceedings only if the Employer agrees to indemnify us against legal costs as envisaged by Clause 3.3.6(b) of the Contract.

Yours faithfully,

8.12   Subcontract notices in general

There are many notices and letters which you must or may have to send to your subcontractors as the subcontract works progress. This applies to all types of subcontractor and, in many respects, these letters and notices parallel those which the architect sends to you under the main contract.

The following is a tabular summary of some of the more important documentation required under NSC/C, NAM/SC and DOM/1. Similar documentation applies under the other standard subcontract forms such as DOM/2, which relates to WCD 98, and IN/SC which is the domestic subcontract form for use with IFC 98.

Content	NSC/C Clause	NAM/SC Clause	DOM/1 Clause
Notification of restrictions in sub-subcontract or contracts of sale	1.7.2		
Notification of divergence or discrepancy between documents etc.	1.8		4.1.6.2, 4.1.7.1
Provide further drawings and details		2.3	
Directions relating to inconsistencies		2.4	4.1.5
Correction of departures from SMM		3.4	
Issue directions regulating the carrying out of the works		5.1, 5.2.1	4.1.1, 4.2.1
Issue directions to open up for inspection and for testing		5.5, 5.6.2	4.3.1
Issue directions regarding work not carried out in a workmanlike manner		5.7	4.3.2.1/.2/.3
Inform that non-complying work can remain			4.3.3.2
Dissent from subcontractor's confirmation of oral instruction			4.4

Content	NSC/C Clause	NAM/SC Clause	DOM/1 Clause
Issue directions regarding work not carried out in a proper or workmanlike manner or not in accordance with health and safety plan			4.3.3A
Obtain for subcontractor any rights or benefits of main contract	1.13	20	22
Notice to commence work on site	2.1		
Direct expenditure of provisional sum for performance specified work			23.3
Notify requirement to amend 'Subcontractor's Statement'			23.5
Notify deficiency in 'Subcontractor's statement'			23.6
Issue directions requiring variations to performance specified work			23.11
Give directions for the integration of performance specified work			23.14
Notification of delay	2.2.1		
Submit particulars and estimate	2.2.3		
Grant extension of time	2.3	12.2, 12.3	11.3
Inform of decision not to revise period for completion	2.3		11.5
Notify shorter period for completion	2.4		11.6
Extend period for completion of subcontract works		12.4	
Fix period for completion of subcontract works			11.7

Content	NSC/C Clause	NAM/SC Clause	DOM/1 Clause
Join with NSC in dispute procedures	2.7		
Notification of failure to complete on time	2.8	13	12.1
Make application for loss and/or expense		14.3	13.4
Pass NSC's notification of practical completion to architect	2.10		
Dissent from subcontractor's notice of practical completion		15.1	14.1
Direct making good defects etc.		15.3	14.3
Issue architect's written instructions to NSC	3.3.1		
Submit objections concerning variations	3.3.2.2		
Inform architect of NSC's disagreement	3.3.2.3		
Confirm instructions in writing	3.3.3.1/.2		
Pass on NSC's request for further information	3.3A.1.1		
Send quotation to quantity surveyor	3.3A.1.2		
Notify NSC of acceptance of quotation	3.3A.3		
Direct that variation is or is not to be carried out	3.3A.4		
Confirm agreement to alter number of days	3.3A.7		
Report result of consultation with NSC	3.4.3		
Inform NSC that non-complying work can remain	3.6.1		

Content	NSC/C Clause	NAM/SC Clause	DOM/1 Clause
Send copy of instruction to take down work to NSC	3.8.1		
Written notice to comply with direction	3.10		
Request architect to specify provision empowering issue of instruction	3.11		
Deliver architect's answer to NSC	3.11		
Join with NSC in dispute procedures	3.11		
Consent to assignment	3.13	24.1	26.1
Consent to subletting	3.14.1	24.2	26.2
Notify whether or not subcontractor's price statement accepted	4.4.2 A2	16.2 A2.2	4.6.3
Notify reasons for non-acceptance	4.4.2 A4.1	16.2 A4.1	4.6.4
Supply amended price statement	4.4.2 A4.1		
Notify acceptance or non-acceptance of NSC's requirements	4.4.2 A7.1.1	16.2 A7.1.1	
Notify acceptance or non-acceptance of adjustment of time for completion	4.4.2 A7.1.2	16.2 A7.1.2	
Daywork vouchers	4.13.3		
Submit application by NSC	4.15.1		
Consent to removal of unfixed materials	4.15.4.1	19.5.1	21.4.5.1
Give written notice of amount of interim payment	4.16.1.1	19.2.3	21.3.2
Discharge payment	4.16.1.1	19.2	21.2
Issue withholding notice	4.16.1.2	19.2.4, 19.8.2.2	21.3.3, 21.9.3
Notify amount of final payment		19.8.2.1	21.9.2

Content	NSC/C Clause	NAM/SC Clause	DOM/1 Clause
Submit NSC's application for payment to architect and quantity surveyor	4.17A		
Join with NSC in dispute proceedings	4.2		
Require architect to operate clause 26.4 of main contract	4.38.1		
Notice by contractor of delay by NSC	4.40		
Confirm satisfaction or non-satisfaction with NSC's tax status	5C.2.1.1		
Confirm production of current tax certificate	5C.2.1.2	18A.2.1.2	20A.2.1.2
Notice of change in contractor's tax status	5C.3.2, 5D.3	18A.3.2, 18B.3	20A.3.2, 20B.3
Pass voucher to Inland Revenue	5C4.3	18A.4.2	20A.4.3
Notification of change of planning supervisor or principal contractor	5E.2.1	18C.1	4A.1
Ensure supply of any development of health and safety plan	5E.2.2	18C.2	4A.2
Endorse joint names policy to recognise subcontractor	6A.1, 6B.1, 6C.1	9A.1, 9B.1, 9C.1.1, 9C.1.2	8A.1, 8B.1, 8C.1
Produce documentary evidence of insurance	6.9	10.3	9.3
Production of insurance policy and receipts	6.9	10.3	9.3
Issue notice requiring remedial measures	6FC.2.1	9FC.2.1	8D.2.1
Notice of default or defaults	7.1.1	27.2.1	29.2.1
Notice of determination	7.1.2, 7.2.4	27.2.2, 27.3.4	29.2.2, 29.3.4
Notice of determination – repeated event	7.1.3	27.2.3	29.2.3

Content	NSC/C Clause	NAM/SC Clause	DOM/1 Clause
Agree reinstatement of employment	7.2.3	27.3.3	29.3.3
Execute adjudication agreement	9A2.3	35A.2.3	38A.2.3
Notice of intention to refer dispute to adjudication	9A4.1	35A.4.1	38A.4.1
Agree adjudicator/apply to nominating body	9A.2.2	35A.2.2	38A.2.2
Notice of arbitration	9B.1.1	35B.1.1	38B.1.1

Four particular situations are of considerable practical importance and are considered in detail here:

• extensions of time
• money claims
• withholding notices
• determination of employment

8.13 Extensions of time

Clauses 2.2–2.7 of NSC/C, clauses 11.2 to 11.10 of DOM/1 and clauses 12.2 to 12.8 of NAM/SC, deal with extensions of the subcontract time and the various provisions are broadly similar; the various letters given and referred to in this section are readily adaptable to the various subcontracts. The first stage in the procedure is the subcontractor's notice to you which he must give 'if and whenever it becomes reasonably apparent that the commencement, progress or completion of [the whole or part of] the subcontract works is being or is likely to be delayed'. It is always difficult to reject a claim as being out of time.

You must, if the subcontractor is nominated under JCT 98, then inform the architect of the subcontractor's notice (see **document 8.9.5**) and you must do this immediately, even if in your view the subcontractor has given insufficient information. Write and ask him for more information (**document 8.13.1**). The subcontractor must keep you up to date with the position and you must pass these details on to the architect.

You cannot grant an extension of time to a nominated subcontractor without the architect's consent but, assuming this is given, you should grant the extension (**document 8.13.2**). Your extensions of time to named or domestic subcontractors are your own concern but obviously, if the reason for the extension is something that gives you the right to an extension of time yourself, you need to take that up with the architect. Note that in giving the extension you must (a) fix a new period for completion, (b) state which relevant events have been taken into account and (c) state the extent, if any, to which regard has been given to any instruction requiring the omission of any work issued since the fixing of the last completion date. The extension must be granted, if reasonably practicable, within 12 weeks (16 weeks under DOM/1) of the date on which you received the subcontractor's notice, reasonably sufficient particulars and estimate. However, if there are less than 12 weeks to the completion date the extension must be granted *before* the completion date. You may need to remind the architect about these time limits (**document 8.13.3**) under NSC/C.

Under NAM/SC you have to grant an extension of time 'so soon as you are able to estimate the length of delay'.

The subcontractor may well press you for a decision, just as you would press the architect for a main contract extension of time in your favour.

Under NSC/C and DOM/1, after practical completion you must notify the subcontractor, not later than 12 weeks (16 weeks under DOM/1) after the date of practical completion, of the result of the final review. You need the architect's written consent where a nominated subcontractor is concerned under NSC/C and your letter must either confirm the date previously fixed or revise it (**document 8.13.4**). At this stage you (and the architect) must take into account all relevant events, whether or not the subcontractor has notified you of them. If, in your view, a relevant event has delayed the subcontractor's progress but he has failed to notify you of it, where a nominated subcontractor is concerned under NSC/C, you should tell the architect when seeking his written consent (**document 8.13.5**).

There is no specific provision for a review under NAM/SC albeit clause 12.4 seems to envisage that a review can take place.

Under NSC/C if the architect refuses his consent to an extension of time, you are not bound to give reasons, even if pressed, but you should notify the subcontractor and send him a copy of the architect's letter (**document 8.13.6**).

The subcontractor is required to use his best endeavours to prevent delay in progress and to do all that may reasonably be required to proceed with the subcontract works.

Note that the subcontractor who is aggrieved by the architect's/your decision or failure to give a decision has a right to go to adjudication, arbitration or litigation. Where a nominated subcontractor is involved he may use your name subject to giving you suitable indemnity and security in any such proceedings. **Document 8.13.7** may be useful in these circumstances.

You must remember whenever considering matters involving extensions of time, that the length of time accepted under the clause 3.3A quotation procedure (NSC/C) cannot be subsequently varied (clause 4.6 under DOM/1).

8.13.1 Letter asking the subcontractor for further information on delays

To the Subcontractor   *Date*
(Copy to the Architect)

Dear Sirs,

[*Heading*]

With reference to your notice of delay dated [*date*], in our
opinion, [which is supported by the Architect (NSC/C)] you
have not provided us with sufficient particulars of the
expected effects of the delays complained of to enable [the
Architect to form an opinion as to whether he should give his
consent to the granting of an extension of time (*NSC/C*)] [us to
consider properly what, if any, extension of time we should
grant (other forms)].

Will you please, therefore, provide us with the following
further particulars as soon as possible: [*specify particulars
needed, e.g.* why the minor variation to the pipe runs in the
boiler room should be estimated to cause a delay of 6 weeks in
the completion of the installation].

Yours faithfully,

8.13.2   Letter granting an extension of time to a subcontractor

To the Subcontractor   *Date*
(Copy to the Architect) (NSC/C only)

Dear Sirs,

[*Heading*]

With reference to your notice of delay dated [*date*] and the
further particulars and estimates provided in your letter
[*date*], [with the written consent of the Architect (*NSC/C*)] we
now grant an extension of time for completion of your
subcontract Works.

The revised period for completion of your Works under the
subcontract is <u>56 weeks</u>, representing an extension of time of 6
weeks.

In fixing this extension of time we have had regard to the
following:
1.   delay by the subcontractor for the electrical
     installation in wiring pumps
2.   variation in pipe runs in the boiler room
[*etc.*]

In granting his/our written consent to this extension, [the
Architect has/we have] had regard to the Architect's
instruction no. [*specify number*] omitting five radiators in
the main hall which, in his/our opinion, reduces the period
required for completion of the subcontract Works by three
days.

Yours faithfully,

8.13.3 Letter reminding the architect of the time limit for granting an extension of time (NSC/C)

To the Architect *Date*

Dear Sir,

[*Heading*]

With reference to the notice of delay submitted by Acme Heating Engineers (1989) Ltd dated [*date*], and the further particulars and estimates submitted by them dated [*date*], which we received on [*date*], we would remind you that Clause 2.3 of the subcontract requires us to grant any appropriate extension of time for completion of their Works, with your written consent, within 12 weeks from the date of receipt of sufficient particulars and estimates. This period, measured from our receipt of the further particulars, expires in one week from today, and we shall therefore be grateful if you will let us have your written consent to the granting of any extension of time which you consider appropriate at last three days before the expiry of that time so that we may have sufficient time to communicate your decision to the subcontractor.

Yours faithfully,

8.13.4   Letter to the subcontractor regarding the final decision on extensions of time

---

To the subcontractor   *Date*
(Copy to the Architect) (NSC/C only)

Dear Sirs,

[Heading]

[In accordance with Clause 2.5 of the subcontract, this is to notify you that the Architect has now given his written consent to the granting to you of a (*NSC/C*)] [In accordance with clause 11.7 of the subcontract we have reviewed the period for completion of your works and we hereby grant you a (*DOM/1*) (*NAM/ SC*)] further extension of the period for completion of your works of <u>two weeks</u>. The period for completion of your Works is therefore <u>62 weeks</u>.

[Addendum where subcontractor fails to complete by revised date:]
Since the date for commencement of your Works was originally fixed as [*date*], the revised period for completion of your Works expired on [*date*]. [We have requested the Architect to certify your failure to complete your Works by that date, and a copy of our notification is enclosed (*NSC/C*)]. In due course we shall notify you of the amount of loss or damage caused to us by your failure to complete your Works within the revised period now fixed for which we shall be seeking reimbursement.

Yours faithfully,

8.13.5 Letter notifying the architect of a previously un-notified ground for the extension of time to a nominated subcontractor (NSC/C)

To the Architect   *Date*

Dear Sir,

[*Heading*]

Concerning your final decision regarding extension of time to Acme Heating Engineers (1989) Ltd, which is due to be communicated to that firm within 12 weeks after certification of practical completion of their Works which expires on [*date*], it has come to our notice that their Works were delayed by a cause for which they have submitted no written notification themselves in that [*specify cause*]. Will you please, therefore, take this into account when reaching your final decision as to the extension of time for which you will give consent under Clause 2.5 of the subcontract.

Yours faithfully,

8.13.6 The architect's refusal of consent to the granting of an extension of time to a nominated subcontractor (NSC/C)

To the Subcontractor   *Date*
(*Copy to the Architect*)

Dear Sirs,

[*Heading*]

With reference to your notice of delay dated [*date*] and the
further particulars provided by you in your letter dated
[*date*], we have to inform you that the Architect has refused
his consent to the granting of an extension of the period for
completion of your Works for the causes notified. A copy of the
Architect's letter is enclosed.

We regret, therefore, that we cannot grant any extension of
time for completion of your Works.

Yours faithfully,

8.13.7 Letter to the nominated subcontractor concerning indemnity and security in respect of dispute proceedings

To the Subcontractor *Date*

Dear Sirs,

[*Heading*]

We have received your letter of [*date*] requesting us to allow you to use our name and to join with you in dispute proceedings concerning the refusal of the Architect to grant an extension of the period for completion of your Works, communicated to you in our letter of [*date*].

As provided in Clause 2.7 of the subcontract, we are willing to allow you to use our name and to join with you in the proceedings subject to receipt from you of the following:
1. an indemnity, in terms to be drafted by agreement between our respective solicitors, in respect of all Costs and other charges arising from the arbitration proceedings;
2. provision of adequate security in respect of such Costs and other charges, also to be agreed in form and amount between our respective solicitors.

Please telephone the writer within the next two days to arrange a meeting at which these matters may be discussed.

Yours faithfully,

8.14   Money claims

These are the subject matter of NSC/C, clause 4.38 to 4.41 and, in addition to the grounds which you enjoy under the main contract, the subcontractor can also claim if regular progress of the subcontract works is materially affected by your acts, omissions or defaults or by those of your other subcontractors and others for whom you are legally responsible.

Your claims against a nominated subcontractor in respect of matters affecting regular progress of the works fall under clause 4.40 (NSC/C) and require you to give written notice to the subcontractor (**document 8.14.1**) and having done so you should endeavour to agree the amount (**document 8.14.2**). Your claims against a nominated subcontractor for delay in completion are covered by clauses 2.8 and 2.9 of NSC/C.

As regards the subcontractor's claims, the first step is that the subcontractor must give you written notice, backed up by supporting detail and information. Once you receive written notice you must send a copy to the architect and request him to ascertain the direct loss and/or expense. The basic letters which may be needed are set out as **documents 8.14.3** to **8.14.5**.

In practice, a meeting between you, the architect, quantity surveyor and subcontractor may well be needed to sort out the claim. It is for the sub-contractor to produce the necessary evidence to enable the architect or quantity surveyor to make the ascertainment; many subcontractors' claims fail because of lack of substantiating evidence.

The provisions of the domestic forms of subcontract, including NAM/SC, are rather simpler. The architect does not get involved. You have to deal with the subcontractor's claim or make a written application to him in the event of his default. The letters given as examples under NSC/C can be adapted appropriately.

8.14.1 Letter to the nominated subcontractor regarding the disturbance to regular progress of the main contract works (NSC/C)

To the Subcontractor   *Date*

Dear Sirs,

[*Heading*]

In accordance with Clause 4.40 of the subcontract, this is to notify you that regular progress of the main Contract Works has been materially affected by [*specify cause, e.g. give details of the manner in which the subcontractor has failed to comply with the agreed programme*].

We shall notify you shortly of the amount of direct loss and/or expense caused to us by this default on your part.

Yours faithfully,

8.14.2 An attempt to agree the amount of direct loss and/or expense caused by the nominated subcontractor (NSC/C)

To the Subcontractor *Date*

Dear Sirs,

[*Heading*]

Further to our letter of [*date*] in which we notified you that regular progress of the main Contract Works has been materially affected by [your failure to comply with the agreed programme], we now enclose details of the direct loss and/or expense incurred by us and caused by your failure.

We shall be glad to receive your agreement to the amount shown. Should you wish to discuss any aspects of the enclosed statement please contact us. Should we not hear from you within the next 7 days we shall assume that the amount is agreed.

We hereby give you the written notice required by Clause 4.16.1.2 of the subcontract that we shall be deducting this amount from the next payment directed to be made to you in the next Architect's Certificate.

Yours faithfully,

Note: The notice in the final paragraph may well be repeated before the payment is made.

8.14.3 Letter to the architect passing on the nominated subcontractor's application for direct loss and/or expense (NSC/C)

To the Architect   *Date*

Dear Sir,

[*Heading*]

We have received an application from Acme Heating Engineers (1989) Ltd stating that they are likely to incur direct loss and/or expense in the execution of their Works by reason of regular progress of their Works being likely to be materially affected for the causes stated in their application, a copy of which is enclosed.

Please let us know if there is any further information you require to enable you to form an opinion as to whether regular progress of their Works is likely to be affected as stated in their application.

Yours faithfully,

8.14.4 Letter to the nominated subcontractor requesting further information (NSC/C)

To the Subcontractor   *Date*

Dear Sirs,

[*Heading*]

With reference to your application dated [*date*] stating that you were likely to incur direct loss and/or expense in the execution of your Works by reason of regular progress of your Works being likely to be materially affected by [*specify cause, e.g.* variations ordered by the Architect] the Architect, in accordance with Clause 4.38.1.2, has now requested that you provide further information regarding the anticipated effect on regular progress sufficient to enable him to form an opinion on the matter. A copy of the Architect's letter is enclosed.

Please provide the further details requested as soon as possible so that we may submit them to the Architect.

Yours faithfully,

Note: A similar letter may be sent in respect of any details of loss and/or expense requested under Clause 13.1.3 of DOM/1.

8.14.5   Letter to the nominated subcontractor conveying the ascertainment made by the quantity surveyor

To the Subcontractor   *Date*

Dear Sirs,

[*Heading*]

Following your application in respect of direct loss and/or
expense claimed by you as arising from the effect upon regular
progress of your works of [*specify*], and the particulars of
loss and/or expense submitted by you with your letter of
[*date*], we have now received the Quantity Surveyor's
ascertainment of the amount due to you. A copy is enclosed.

Please let us know as soon as possible whether you accept this
ascertainment so that we may arrange for payment to be included
in the next Interim Certificate due on [*date*].

Should you have any queries concerning the ascertainment
please let us know at once so that we may, if necessary, arrange
a meeting with the Quantity Surveyor.

Yours faithfully,

8.15   Withholding notices

There is no longer any right of set-off set out in any of the subcontracts as there used to be under earlier editions of the subcontracts allied to the JCT forms prior to the 1998 contracts. This has been replaced by the right of the paying party to give written notice specifying any amount proposed to be withheld and/or deducted from the amount otherwise due to the payee.

In order to be effective a notice of intent to withhold an amount must be given no later than five days before the final date for payment and it must set out the ground or grounds for such withholding and/or deduction and the amount of the withholding and/or deduction attributable to each ground.

Where a written notice is not given, the full amount certified by the architect, in the case of NSC/C, or the full amount set out in the notice that you are required to give under clause 21.3.2 of DOM/1 or 19.2.3 of NAM/SC, is payable to the subcontractor. Failure to make payment in full in these circumstances results in the subcontractor having the right, after giving seven days notice, to suspend performance of the subcontract or to inform you that he disputes the amount withheld and issue you with a notice of adjudication.. **Document 8.15.1** is a suggested form of withholding notice.

8.15.1 Withholding notice

SPECIAL DELIVERY

To the Subcontractor   *Date*

Dear Sirs,

[*Heading*]

In accordance with Clause [4.16.1.2 (*NSC/C*) or 21.3.3 (*DOM/1*)
or 19.2.4 (*NAM/SC*)] of the Subcontract we hereby give you
notice that it is our intention to withhold from the sums [set
out in the Architect's Interim Certificate dated [*date*] (*NSC/
C*) *or* set out in our notice dated [*date*] under Clause [21.3.2
(*DOM/1*) *or* 19.2.3 (*NAM/T*)]] the amount of our claim for direct
loss and/or expense actually incurred by us at today's date in
respect of [*specify breach or default*] and which we have
quantified in detail as you will see from the attached Schedule
and supporting documentation. [*The costs and charges actually
incurred should be itemised, e.g. scaffolding etc., and back-
up evidence provided.*]

Yours faithfully,

8.16 Determination

This section should be read in conjunction with Chapter 7 and there are four situations to be considered:

- Default by subcontractor
- Insolvency of subcontractor
- Determination by subcontractor
- Determination under main contract

In all cases the complexities are such that legal advice is desirable, but there are some basic notices and letters which must be sent and the documents in Chapter 7 may be used as a guide, with appropriate alterations.

Chapter 9
Settlement of Disputes

9.01 Introduction

The dispute resolution aspects of construction contracts changed dramatically with the introduction of the statutory right to adjudication in the Housing Grants, Construction and Regeneration Act 1996 (the HGCRA) which came into force on 1 May 1998.

In all JCT contracts entered into prior to this date, the only means of settling disputes, other than the sensible way of negotiation, was to refer the dispute to arbitration. This was, in most situations, not available until after practical completion of the contract works. There was the option of striking out the arbitration clause and placing any dispute in the hands of the court, but wisely this was not done in the vast majority of construction contracts.

With the coming of the HGCRA, the JCT amended the dispute resolution procedures and introduced adjudication into the contracts. The JCT also decided that arbitration would no longer be the only forum for the final resolution of disputes that the parties could not resolve themselves, and provided alternative standard clauses which allow the parties to choose either arbitration or litigation for this purpose.

In this chapter, I deal with the procedures necessary to instigate adjudication and arbitration. If you enter into a contract in which the arbitration clause is deleted and you are unable to resolve a dispute by agreement either before or after adjudication, you will be in the unfortunate position of having to go to litigation. If this happens you would be foolhardy not to engage the services of a solicitor to guide you. This is beyond the realms of this book.

The JCT contracts do no more than require that you 'give notice to the other Party' when notifying your intent to refer a dispute to adjudication. In the examples in this chapter I have followed the same practice as in earlier chapters and annotated each with the words 'By Special Delivery'. In practice these notices will in all likelihood be sent by fax and followed up by post.

9.02   Adjudication generally

Every JCT contract now includes adjudication provisions in order to comply with the requirements of the HGCRA. There is now no longer the requirement to wait until practical completion before you can refer a dispute to a third party; this can be done at any time.

Adjudication does not settle a dispute. All it does is provide what is graphically described as a 'temporarily binding' decision. It gives the parties the opportunity to receive the adjudicator's decision on their respective rights and obligations under the contract. The adjudicator's decision does however have to be complied with. It will be enforced by the courts if the party that is ordered to pay does not do so. Historically, the contractor, or a subcontractor, has had to make the decision to pursue a dispute into arbitration if he feels that he has been hard done by during the course of the contract. This has now changed and the position of the parties is often reversed. It is the party that has been required to pay money by an adjudicator's decision that has to decide to take the dispute further if it feels that the decision was wrong. This could well be the employer rather than the contractor.

Any dispute can be referred to adjudication at any time. It has, however, to be a dispute that arises 'under the contract'. Allegations of fraudulent misrepresentation, for example, are a tort. This is not a matter that arises under the contract. All is not lost however. There is no longer any restriction on a reference to arbitration and this can be done during the course of the contract.

Most adjudications result from allegations that money that should have been paid is being held onto by the paying party. The mere reference to adjudication often brings the paying party to its senses and payment is made. Often payment is made during the course of the adjudication before the adjudicator produces his decision.

A proportion of adjudications relate to payment during the course of the contract or final accounts where the sums in dispute are large and arguments about extensions of time and loss and expense are involved. There is little merit in this sort of instance in standing on ceremony and insisting on a decision in the 28 days allowed unless time is extended. If the adjudicator is forced to reach a decision in that time scale it is almost inevitable that you, or the other party, will be unhappy with the result. Adjudication is after all only temporarily binding. It is almost inevitable that the losing party will want to have another go if they feel that a proper consideration has not been given to the matters in dispute. Far better to allow sufficient time in these circumstances for a proper investigation and for you and the other party to provide the adjudicator with a detailed submission. In that way there is a chance that the adjudicator's decision will be accepted or form the basis of a negotiated settlement, and that all the time and expense of subsequent arbitration or litigation proceedings can be avoided.

You may find that the other party or the adjudicator do not see things quite in the same way as you do and you cannot get an extension of time agreed. Remember that adjudication is a right not an obligation. You do not have to refer a dispute to adjudication. You can take it to arbitration or litigation. It may be that you would prefer to retain some control and refer it immediately to arbitration.

It is easy to carry on with the contract in the hope that you will be able to sort it all out after the job is finished. That was how things were before the statutory right to adjudication and things could turn out to be very difficult in some instances. The secret is, however, not to allow a dispute to escalate to that level and always to consider your options at the earliest possible stage. An early reference to adjudication may prevent a lot of heartache later.

Any further discussion of the pros and cons of adjudication and the procedures that apply to the adjudication process are outside the scope of this book. You are referred to *Construction Adjudications* by John Riches and Christopher Dancaster (Lloyds of London Press, 1999) for further details.

9.03 Appointing an adjudicator

Adjudication is a very quick process. The appointment of an adjudicator is equally quick. If you are the party that wishes to have a dispute referred to adjudication, you are entitled to have an adjudicator in place within seven days of you issuing a notice of your intention to refer that dispute to adjudication ('the notice of adjudication') to the other party. You do this by means of 'the referral' which is sent to the adjudicator accompanied by a summary of the contentions relied on, a statement of the relief or remedy sought and any material that you wish the adjudicator to consider.

The appointment of the adjudicator results from 'the referral'. There is no need to agree the adjudicator's identity. The referral commences the adjudication process and unless the time is extended, the adjudicator has to produce his decision within 28 days.

You may have agreed the identity of the adjudicator and his name may be inserted in the contract. In this event as soon as you have given the employer the notice of adjudication (**document 9.03.1**) all you have to do is to prepare your referral notice and accompanying documents and send them to the named adjudicator (**document 9.03.2**). A copy of this notice and the accompanying documents must be sent to the employer.

The naming of an adjudicator in the contract is undesirable. You have no idea at that time of the nature of any dispute that may arise under the contract. Given the fact that the normal dispute relates to payment, it is likely that a quantity surveyor will be named. He may be very good if the dispute relates to allegations of underpayment but he may be totally unsuitable to deal with a dispute which, for example, relates to compliance with the specification.

It is important to remember that adjudication is a different skill from arbitration. It may have many similar characteristics and need similar skills in interpreting contracts and the obligations that parties have undertaken, but an arbitrator is often chosen for his skill as an arbitrator, and familiarity with the technical aspects of the dispute, while very important, can often be secondary. In adjudication a knowledge of and experience in adjudication procedure is vital but equally vital is an understanding of the technical aspects of the matters in dispute. In adjudication there is no time for the adjudicator to find out about such matters; he has to have them at his fingertips.

My advice would be to demur from accepting a named adjudicator in the contract on the basis that the benefits of the appointment of an individual suitable for the particular dispute in question will outweigh the slight delay of getting an adjudicator appointed when that dispute arises.

When there is no adjudicator named in the contract your immediate problem on issuing the notice of adjudication is to find a suitable name. You do have the opportunity to name someone in your notice of adjudication and it is possible to seek suitable names from one or more of the organisations offering services as nominating bodies. There is a list of four such bodies in the appendix to the contract and in seeking names you are not restricted to the body identified in the appendix; you can use one of the three that are struck out or even another nominating body if you feel that a more suitable name for the particular dispute may result. One body worth thinking about in this respect is the Chartered Institute of Arbitrators. This may seem a strange suggestion to some as an adjudicator nominating body but it does have a panel of construction industry adjudicators of a wide range of disciplines and applies very rigorous standards to those on its panel. You then include a list of names in your notice of adjudication for the employer to choose one from. This may delay the appointment of the adjudicator as you will have to give the employer at least 24 hours to agree a name. If he does not do this you have to apply to the adjudicator nominating body named in the contract for a nomination and that nominating body may have difficulty in making that nomination within the required seven days of your adjudication notice. **Document 9.03.3** is an example of a notice of adjudication with suggested names of adjudicators.

There is no obligation on you in the JCT contracts to agree the name of the adjudicator with the employer and it may well be that the most appropriate action is to apply immediately to the adjudicator nominating body in the appendix.

Document 9.03.4 is an example of the form that the Royal Institution of Chartered Surveyors requires you to fill in when applying for a nomination. You will note that the RICS requires a copy of the notice of adjudication to accompany this form. This form gives you the opportunity to indicate the particular expertise of the adjudicator, for example 'a quantity surveyor with experience of dealing with extension of time claims', and nominating bodies are generally very careful in trying to find a nominee who fits in with the stated requirements. The RICS among others has a very stringent assessment process before including anyone on its list and that list is not restricted to surveyors.

Document 9.03.2 will also be used in this instance to refer the dispute to the adjudicator. This has to be given to the adjudicator by actual delivery, by fax, by special delivery or by recorded delivery. If by fax it has to be followed up by hard copy by first class post or actual delivery. In practical terms, where the documentation is in any way extensive sending it by fax can be a long process and the adjudicator may not be best pleased if his fax machine is blocked for a long period, especially as he will received hard copy the next day. Once again a copy of the referral notice and the accompanying documents must be sent to the employer.

The contract requires that the JCT Adjudication Agreement (**document 9.03.5**) is completed by you, the employer and the adjudicator but any failure to do this does not invalidate the decision of the adjudicator.

The adjudicator is now appointed and may set his own procedure. The only specific requirement in the contract is that 'the party not making the referral' (the employer, when you instigate the adjudication process) has the option to send a written statement of his contentions to the adjudicator within seven days of the date of the referral. He has to send a copy to you as well. Other than this requirement the procedure for the adjudication is in the hands of the adjudicator.

He should give you the opportunity to reply to the submission made by the employer and it is often the case that an adjudicator will seek the opportunity to meet with the parties and this is an excellent opportunity to overcome some of the problems of the limited time scale in clarifying any points to which you feel that the adjudicator's attention ought to be brought.

9.03.1   Notice of adjudication where adjudicator named in contract

BY FAX AND ACTUAL/SPECIAL/RECORDED DELIVERY

To the Employer   *Date*
Copy to Architect [out of courtesy (except WCD 98)]

Dear Sir,

[Heading]

We acknowledge receipt of your cheque in respect of Interim
Certificate no. 6 dated *[date]*.

The amount of this cheque is a total of £15,000.00 less than the
total on the Interim Certificate and we note that this equates
to the deduction of Liquidated and Ascertained Damages in
respect of three weeks delay which has not been accepted by the
Architect.

You will be aware that we applied to the Architect on *[date]* for
an extension of time amounting to six weeks on the basis of the
failure by the Architect to comply with clause 5.4.1 of the
Information Release Schedule (JCT 98). We maintain the view
that six weeks is the proper extension of time resulting from
the Architect's default in this respect.

We appear to have no alternative other than to refer this
dispute to adjudication and hereby give the notice required by
Clause [41A.4.1 *(JCT 98)* or 9A.4.1 *(IFC 98)* or 39A.4.1 *(WCD 98*
or 8.1 D4.1 *(MW 98)*].

Our referral notice will be issued tomorrow to Mr *[xxx]*, the
adjudicator named in the contract.

Yours faithfully,

9.03.2   Referral notice

BY ACTUAL/SPECIAL/RECORDED DELIVERY

To the Adjudicator   *Date*
Copy to Employer

Dear Sir,

[Heading]

We are in dispute with the Employer in respect of an extension
of time granted to us by the Architect.

We hereby refer this dispute to you as the adjudicator named in
the Contract. [We are notified of your nomination as
adjudicator by the Royal Institution of Chartered Surveyors
and we hereby refer this dispute to you.]

We enclose the details as required by Clause [41A.4.1 (*JCT 98*)
or 9A.4.1 (*IFC 98*) *or* 39A.4.1 (*WCD 98*) *or* 8.1 D4.1 (*MW 98*)]
comprising the following:

- A summary of the contentions on which we rely
- A statement of the relief and remedy that we seek

We also enclose a file of documents which includes a copy of the
contract and copies of the correspondence relating to the
delay and the inadequate extension of time granted by the
Architect.

We have copied all this material to the Employer.

We look forward to receiving any directions that you may issue
and your decision within 28 days of this referral.

Yours faithfully,

9.03.3 Notice of adjudication where adjudicator not named in contract

---

BY FAX AND ACTUAL/SPECIAL/RECORDED DELIVERY

To the Employer   *Date*
Copy to Architect [out of courtesy (except WCD 98)]

Dear Sir,

[Heading]

We acknowledge receipt of your cheque in respect of Interim Certificate no. 6 dated *[date]*.

The amount of this cheque is a total of £15,000.00 less than the total on the Interim Certificate and we note that this equates to the deduction of Liquidated and Ascertained Damages in respect of three weeks delay which has not been accepted by the Architect.

You will be aware that we applied to the Architect on *[date]* for an extension of time amounting to six weeks on the basis of the failure by the Architect to comply with clause 5.4.1 of the Information Release Schedule (JCT 98). We maintain the view that six weeks is the proper extension of time resulting from the Architect's default in this respect.

We propose to refer this dispute to adjudication and hereby give the notice required by Clause [41A.4.1 *(JCT 98)* or 9A.4.1 *(IFC 98)* or 39A.4.1 *(WCD 98)* or 8.1 D4.1 *(MW 98)*]

We propose Mr [*xxx* of (address)] or Mr [*yyy* of (address)] as adjudicator in this dispute. If we do not receive your agreement to him acting as Adjudicator by 4PM tomorrow we shall apply to the nominator named in the appendix to the contract.

Yours faithfully,

9.03.4

Form AS2A(March 1999)

Dispute Resolution Service
Surveyor Court
Westwood Way
Coventry CV4 8JE
Tel +44 (0) 171 222 7000
**   +44 (0) 1203 694757**
Fax +44 (0) 171 334 3802
Email drs@rics.org .uk

APPLICATION FOR THE NOMINATION OF AN ADJUDICATOR BY THE ROYAL INSTITUTION OF CHARTERED SURVEYORS

You are encouraged to type all details

I/We **hereby request the** **Royal Institution of Chartered Surveyors nominate an Adjudicator to act in the case described overleaf:** I/we accept that the nomination will be made through the RICS Dispute Resolution Service and this is the basis upon which this application is submitted to you and upon which the application will be entertained. (Please note section **(D)** of this application form - overleaf).	
Applicant/Claimant (full names and address)	
Applicant's Representative (name, address, telephone number and reference. State whether solicitor, surveyor, company official) ie person or firm to whom communication should be sent.	
Other Party/Respondent (full names and address)	
Other Party's Representative (name, address, telephone number and reference. State whether solicitor, surveyor, company official) ie person or firm to whom communication should be sent.	

Continued Overleaf

(A) Nature of Dispute

To enable the RICS to nominate a suitable adjudicator you will need to identify the precise nature of the dispute. If you do not it may delay the settlement of the dispute. Please ensure that **FULL** details of the matters to be adjudicated are described here, and enclose a copy of the **NOTICE OF ADJUDICATION** with this application. (Continue on a separate sheet if necessary).

(B) Qualifications of the adjudicator

If you think there are specific qualities or expertise that in your opinion would be required of the person to be nominated, please list them here. (Continue on a separate sheet if necessary). It is emphasised that while careful consideration will be given to any representations, the RICS will have the final decision as to who shall be nominated.

(C) Conflicts of Interest

If you think there are any persons who should not be considered for nomination, please state their names here, together with full reasons supporting your views. (Continue on a separate sheet if necessary). It is emphasised that while careful consideration will be given to any representations, the RICS will have the final decision as to who shall be nominated.

(D) Agreement to Refer (delete as appropriate)

Is this application made in accordance with adjudication provisions contained in the contract between the parties? **YES/NO** *(If yes, please provide a copy of the signed contract with this application).*

Does the contract between the parties refer specifically to the <u>President</u> of the RICS as the nominating body? **YES /NO** *(If no, the nomination will be made by the RICS Dispute Resolution Service on behalf of the Institution).*

Is this application made in accordance with the Scheme for Construction Contracts? **YES/NO**

(E) Does the contract specify the adjudication provisions which are to apply?

If yes, provide a copy of the relevant contractual provision(s)

(F) Fees

A fee of £235.00 (inclusive of VAT) must accompany all applications for the nomination by the Royal Institution of Chartered Surveyors. The fee is non-returnable whether or not the Institution makes the appointment (e.g. if the matter is settled by agreement).

I/We **enclose** a cheque for £235.00 made payable to the RICS Business Services Limited.

I/We **undertake** to be responsible for payment of the reasonable professional fees and costs of the person nominated, including any fees and costs where a negotiated settlement is reached before the decision of the adjudicator is made.

Signed ....

Dated ....

To be returned to:

The Dispute Resolution Service, Surveyor Court, Westwood Way, Coventry, CV4 8JE.
Tel 0171 222 7000 - (or Local Calls : 01203 694757) - Fax 0171 334 3802 - Email drs@rics.org.uk

RICS Business Services Ltd Registered in England No.1526902 Registered Office: Surveyor Court Westwood Way Coventry CV4 8JE

9.03.5

JCT Adjudication Agreement

JCT

This Agreement

is made on the _____ day of _____ 19 _____

BETWEEN ('the Contracting Parties')

Insert names and addresses of the Contracting Parties

(1) _____

_____

_____

_____

(2) _____

_____

_____

_____

and ('the Adjudicator')

Insert name and address of Adjudicator

_____

_____

_____

_____

JCT Adjudication Agreement

Whereas

the Contracting Parties have entered into a
*Contract/Sub-Contract/Agreement (the 'contract') for

Brief description of the works/the sub-contract works

_____

_____

_____

_____

_____

on the terms of

Insert the title of the JCT Contract/Sub-Contract/Agreement and any amendments thereto incorporated therein

_____

_____

_____

in which the provisions on adjudication ('the Adjudication Provisions') are set out in

clause _____

And Whereas

a dispute or difference has arisen under the contract which the Contracting Parties wish
to be referred to adjudication in accordance with the said Adjudication Provisions.

*Delete as appropriate.

Adj1/2

JCT Adjudication Agreement

Now it is agreed that

Appointment and acceptance

1   The Contracting Parties hereby appoint the Adjudicator and the Adjudicator hereby accepts such appointment in respect of the dispute briefly identified in the attached notice.

Adjudication Provisions

2   The Adjudicator shall observe the Adjudication Provisions as if they were set out in full in this Agreement.

Adjudicator's fee and reasonable expenses

3   The Contracting Parties will be jointly and severally liable to the Adjudicator for his fee as stated in the Schedule hereto for conducting the adjudication and for all expenses reasonably incurred by the Adjudicator as referred to in the Adjudication Provisions.

Unavailability of Adjudicator to act on the referral

4   If the Adjudicator becomes ill or becomes unavailable for some other cause and is thus unable to complete the adjudication he shall immediately give notice to the Contracting Parties to such effect.

Termination

5   ·1   The Contracting Parties jointly may terminate the Adjudication Agreement at any time on written notice to the Adjudicator. Following such termination the Contracting Parties shall, subject to clause 5·2, pay the Adjudicator his fee or any balance thereof and his expenses reasonably incurred prior to the termination.

      ·2   Where the decision of the Contracting Parties to terminate the Adjudication Agreement under clause 5·1 is because of a failure by the Adjudicator to give his decision on the dispute or difference within the time-scales in the Adjudication Provisions or at all, the Adjudicator shall not be entitled to recover from the Contracting Parties his fee and expenses.

JCT Adjudication Agreement

As Witness

the hands of the Contracting Parties and the Adjudicator

Signed by or on behalf of:　　the Contracting Parties

(1)　　_____

in the presence of　　_____

(2)　　_____

in the presence of　　_____

Signed by:　　the Adjudicator　　_____

in the presence of　　_____

Schedule

Fee　　The lump sum fee is £_____

or

The hourly rate is　£_____

9.04   Arbitration generally

Arbitration is the last resort. When a dispute arises, the first and preferred step is negotiation. Even after a dispute has been referred to adjudication, if you or the employer or a subcontractor are unhappy with the result, it must be sensible to avoid the time and cost of referring the matter to an arbitrator or the courts and seek to agree a way forward with the other party.

When deciding to invoke the arbitration clause you must bear in mind that the eventual outcome of the proceedings is always uncertain. The other party will be in the same position as you. Negotiations will only break down in the absence of goodwill to settle the dispute on both sides or where one or both parties believe that the other party is not prepared to concede enough. Both parties are pretty sure that they are right and are therefore going to make every effort to succeed in the arbitration proceedings. You should always bear in mind that the other side might win.

In all cases you will incur some costs, only part of which are recoverable even if you are successful; if you are unsuccessful, you will probably have to pay a substantial proportion of the employer's costs as well as the whole of your own.

Many arbitrations are settled before the actual hearing or award, but going to arbitration is a serious step and you are probably not the best judge of the strength and merits of your case. You should, therefore, always seek proper professional and/or legal advice before taking action; and certainly you should be reluctant to get involved in a full-scale conventional arbitration which is akin to legal proceedings.

9.05   Appointing an arbitrator

Each of the JCT contracts lays down the procedure for the appointment of an arbitrator. The first step is to write to the employer (*not* the architect) sending him a written notice of arbitration identifying the dispute and requiring him to agree to the appointment of an arbitrator. There is no specific requirement to suggest any names of prospective arbitrators but as the arbitration is being commenced by you, little is likely to happen if you do not do this. The notice should also state that failing agreement on the appointment within 14 days, the appointing body named in the contract will be asked to make an appointment.

It is usual to list the names of three persons who would be acceptable to you. However, the arbitrator must be independent and impartial and must have no subsisting relationship with you, the employer or any other person involved.

If you are in difficulty lists of suitably qualified people can be obtained from:

The Secretary General
Chartered Institute of Arbitrators
24 Angel Gate
City Road
London EC1V 2RS
Telephone 020 7837 4483
Fax 020 78374185

So that appropriately-qualified people may be suggested to you, it is best to give the Institute an indication of the nature of the dispute and the sum of money involved. Before taking any formal step, must be sure in your own mind that there is a genuine dispute or difference between the parties which is capable of resolution by a third party.

Document 9.05.1 is a typical letter asking the employer to concur in the appointment of an arbitrator.

Some employers think that to concur in the appointment of an arbitrator is a sign of weakness. This attitude merely displays a fundamental misunderstanding of the nature of arbitration; indeed, the employer's prompt concurrence in the appointment of an arbitrator could well indicate confidence in his own case.

The employer may, of course, object to all the names you propose and may have good reasons for so doing. If he suggests alternatives these should not be rejected out of hand, but should be considered carefully on their merits. Every effort should be made to agree on an arbitrator acceptable to both parties since it must be remembered that an appointment by a third party, such as the president of the RIBA, is by law binding on both parties and the person so appointed will be almost impossible to remove, except by his own consent – to which he may attach a considerable fee.

The time limit of 14 days runs from the date of your initial letter. If the employer does not concur in the appointment of an arbitrator within that time, you may apply to the president of whichever body is named in the contract as the appointor, to appoint an arbitrator. The three possible appointors are the president or vice-president of the Royal Institute of British Architects, Royal Institution of Chartered Surveyors or the Chartered Institute of Arbitrators. The relevant addresses are:

The Legal Adviser
Royal Institute of British Architects
66 Portland Place
London W1N 4AD

The Dispute Resolution Service
Royal Institution of Chartered Surveyors
Surveyor Court
Westwood Way
Coventry CV4 8JE

The Secretary General
Chartered Institute of Arbitrators
24 Angel Gate
City Road
London EC1V 2RS

You will receive a letter from the relevant body in return, together with a note on the procedure, and a form of application which must be completed and returned to the appointing body.

Document 9.05.2 is the Form of Application to the President of the RICS. A similar document is used by the RIBA and the Chartered Institute of Arbitrators.

Document 9.05.2 can be amended to include a joint application by having it signed by both you and the employer. If the employer suggests that the president should make the appointment, and you agree, the form will be completed accordingly. However, should the employer not do this you will make the application in your own name.

Once the arbitrator is appointed, the procedure to be followed is governed by the JCT 1998 edition of the Construction Industry Model Arbitration Rules. Any discussion of the procedure is outside the scope of this book, and you are referred to *Construction Arbitration* by Vincent Powell-Smith, John Sims and Christopher Dancaster (Blackwell Science, 1998).

9.05.1 Letter requesting concurrence in the appointment of an arbitrator

To the Employer *Date*

Dear Sir,

[*Heading*]

You will be aware that the negotiations which followed the decision of the Adjudicator have failed to resolve the question of the appropriate extension of time that we should have been granted.

In accordance with Clause [41.B.1.1 (*JCT 98*) *or* 39.B.1.1 (*WCD 98*) 9B.1.1 *or* (*IFC 98*) *or* 9.3E2.1 (*MW 98*)] we hereby give you notice of arbitration in respect of this dispute and in accordance with Rule 2.1 of the JCT 1998 edition of the Construction Industry Model Arbitration Rules we require you to agree the appointment of an Arbitrator.

The following are the names of three persons who will be acceptable to us as arbitrator and who have no subsisting connection of any kind with ourselves or, so far as we are aware, with any other person concerned:
[(*Names as appropriate*)]

Failing your agreement to the appointment of one of the above-named persons within 14 days of today's date or your proposals within that time of an arbitrator acceptable to us, we shall apply to the President of [the Royal Institute of British Architects *or* the Royal Institution of Chartered Surveyors *or* the Chartered Institute of Arbitrators (*as appropriate*)] for an arbitrator to be appointed by him within the terms of the Contract.

Yours faithfully,

9.05.2

Form AS2 (May 1999)

Dispute Resolution Service

APPLICATION FOR APPOINTMENT / NOMINATION OF AN ARBITRATOR / INDEPENDENT EXPERT BY THE PRESIDENT OF THE ROYAL INSTITUTION OF CHARTERED SURVEYORS

(Other than commercial property rent review and Agricultural Holdings Act cases)

You are encouraged to type or print all details

I / We	hereby request the President of the

Royal Institution of Chartered Surveyors to [appoint / nominate] an [Arbitrator / Independent Expert]

(delete as appropriate) to act in the case described overleaf
OR
The application for the appointment / nomination of an Arbitrator / Independent Expert to which the following details refer was made in a letter from:

dated: ref:

Applicant / Claimant (full names and address)	
Applicant's Representative (name, address, telephone number and reference. State whether solicitor, surveyor, company official) ie person or firm to whom communication should be sent	
Other Party / Respondent (full names, address and reference)	
Other Party's Representative (name, address, telephone number and reference. State whether solicitor, surveyor, company official) ie person or firm to whom communication should be sent	

Conflicts of Interest
Please state below the names of any persons who, in your view, should not be considered for appointment / nomination (continue on a separate sheet if necessary). It is emphasised that, while the President will give careful consideration to any representations, he will reach his own decision as to who shall be appointed / nominated. Please note that objections will not be entertained unless full reasons are given.

Nature of Dispute
(including approximate sum of money in dispute, location of works and/or address of premises, if relevant; continue on a separate sheet if necessary).

Agreement to Refer
Which clause in the contract or other relevant agreement gives the President power to make this appointment/ nomination? ....

If this application relates to a building dispute, do the Construction Industry Model Arbitration Rules (CIMAR) apply? Yes/No *(delete as appropriate)*.

If CIMAR do apply, has an arbitrator already been appointed in a related dispute? Yes/No. Name ....

A copy of the duly executed (not draft) contract/agreement/Court Order which purports to give rise to the appointment should accompany this application. **Whilst the President may have regard to the contract/agreement supplied, this application form is the contract between the applicant and the President and he will rely entirely upon the information contained herein.**

We accept that in some circumstances the appointment will be made by the President through one of his Vice-Presidents or duly appointed agents and this is the basis upon which the application is submitted to you and upon which the application will be entertained. We accept that in special circumstances (to be decided by the President) it may be inappropriate for the President to effect the appointment and in these circumstances the appointment may be effected by a Vice President in his own name.

Fees
A fee of £235.00 (inclusive of VAT) which is solely for administrative costs must accompany all applications for appointment / nomination by the President. The fee is non-returnable whether or not the President makes the appointment (eg if the matter is settled by agreement).

I / We enclose a cheque for £235.00 (made payable to the RICS Business Services Ltd).

I/We undertake to ensure that the reasonable professional fees and costs of the Surveyor appointed/ nominated are paid, including any fees and costs arising where a negotiated settlement is reached before the award/determination is taken up.

Signed ....

Dated ....

To be returned to : The Dispute Resolution Service, RICS, Surveyor Court,
Westwood Way, Coventry, CV4 8JE
Tel: 0171 222 7000 (or Local Calls: 01203 694757) Fax: 0171 334 3802

9.06 Litigation

If, in the employer's wisdom, he has decided to delete the arbitration clause and you have concurred in this, your only resort in the case of failed negotiations is to refer the dispute to the court. In this event your only option, except perhaps in the case of a dispute where the sums are very small, is to consult your solicitor.

Index